Christoph Thiel

Quantum information and quantum correlations of single-photon emitters

AF061192

Christoph Thiel

Quantum information and quantum correlations of single-photon emitters

Higher-order interference effects in quantum imaging and projection of entangled quantum states

Südwestdeutscher Verlag für Hochschulschriften

Impressum / Imprint
Bibliografische Information der Deutschen Nationalbibliothek: Die Deutsche Nationalbibliothek verzeichnet diese Publikation in der Deutschen Nationalbibliografie; detaillierte bibliografische Daten sind im Internet über http://dnb.d-nb.de abrufbar.
Alle in diesem Buch genannten Marken und Produktnamen unterliegen warenzeichen-, marken- oder patentrechtlichem Schutz bzw. sind Warenzeichen oder eingetragene Warenzeichen der jeweiligen Inhaber. Die Wiedergabe von Marken, Produktnamen, Gebrauchsnamen, Handelsnamen, Warenbezeichnungen u.s.w. in diesem Werk berechtigt auch ohne besondere Kennzeichnung nicht zu der Annahme, dass solche Namen im Sinne der Warenzeichen- und Markenschutzgesetzgebung als frei zu betrachten wären und daher von jedermann benutzt werden dürften.

Bibliographic information published by the Deutsche Nationalbibliothek: The Deutsche Nationalbibliothek lists this publication in the Deutsche Nationalbibliografie; detailed bibliographic data are available in the Internet at http://dnb.d-nb.de.
Any brand names and product names mentioned in this book are subject to trademark, brand or patent protection and are trademarks or registered trademarks of their respective holders. The use of brand names, product names, common names, trade names, product descriptions etc. even without a particular marking in this work is in no way to be construed to mean that such names may be regarded as unrestricted in respect of trademark and brand protection legislation and could thus be used by anyone.

Verlag / Publisher:
Südwestdeutscher Verlag für Hochschulschriften
ist ein Imprint der / is a trademark of
OmniScriptum GmbH & Co. KG
Heinrich-Böcking-Str. 6-8, 66121 Saarbrücken, Deutschland / Germany
Email: info@svh-verlag.de

Herstellung: siehe letzte Seite /
Printed at: see last page
ISBN: 978-3-8381-1163-6

Zugl. / Approved by: Erlangen, FAU, Diss., 2009

Copyright © 2009 OmniScriptum GmbH & Co. KG
Alle Rechte vorbehalten. / All rights reserved. Saarbrücken 2009

Contents

1 Introduction   5

2 Matter-field interaction of single-photon emitters   11
    2.1 Basic system (far-field detection) 12
       2.1.1 Description of the measurement cycle 12
       2.1.2 Quantum paths and loss of which-way information 13
       2.1.3 Geometry of the setup 14
       2.1.4 Two-level systems and the atomic projection operator 15
       2.1.5 Time evolution of the atomic projection operator 16
    2.2 Correlation functions and detection probabilities 19
       2.2.1 The electric field 19
       2.2.2 Intensity correlation function of first order 19
       2.2.3 Intensity correlation functions of higher orders 20
       2.2.4 Interpretation of intensity correlations for a system of atomic emitters 21
       2.2.5 Correlation functions and the detection operator (two-level system) 22

3 Quantum interferences in the light of single-photon emitters   25
    3.1 Quantum interferences in the fluorescence light of uncorrelated single-photon emitters 25
    3.2 The physics behind the concept 28
    3.3 Quantum interferences in the regime of dipole-dipole interaction 31
    3.4 A proof of quantum correlations using Bell inequalities 34
       3.4.1 Derivation of Bell inequalities for single-photon emitters 34
       3.4.2 Violating Bell inequalities by position correlations 36
    3.5 Alternative detection scheme based on optical fibers 38

4 Quantum imaging using single-photon emitters   41

4.1		Introduction to quantum imaging and Abbe's criterion of resolution	41
	4.1.1	Example 1: N00N-state lithography	42
	4.1.2	Example 2: N00N-state microscopy	44
4.2		A new ansatz for quantum imaging using incoherent photons	46
	4.2.1	Model for quantum imaging based on Nth order correlations	46
	4.2.2	Examples for $N = 2$ and $N = 4$ emitters	49
	4.2.3	Quantum microscopy	50
	4.2.4	Experimental feasibility and conclusions	53
4.3		Quantum imaging of an aperture with sub-classical resolution	55
	4.3.1	Description of the experimental configuration	55
	4.3.2	Derivation of the disturbed electric field	56
	4.3.3	Quantum imaging of a rectangular aperture with sub-classical resolution for $N = 2$ emitters	58
	4.3.4	Quantum imaging of a rectangular aperture with sub-classical resolution for $N = 4$ emitters	60
	4.3.5	Quantum imaging of a grating with M slits and sub-classical resolution for $N = 2$ emitters	61
4.4		Conclusions: a comparison with experiment	63

5 Quantum state engineering 65

5.1		Introduction to quantum state engineering	66
	5.1.1	Atom-photon entanglement	66
	5.1.2	Description of the physical system employing emitters with Λ-level structure	66
	5.1.3	A measurement scheme based on *projection*	68
	5.1.4	Example: engineering 2-qubit quantum states	70
5.2		Generation of arbitrary symmetric Dicke states in remote qubits	72
	5.2.1	Introduction to multi-partite entanglement	72
	5.2.2	Symmetric Dicke states of an N-qubit compound system	73
	5.2.3	Description of the physical system	73
	5.2.4	Preparation of symmetric 3-qubit Dicke states	75
	5.2.5	Preparation of symmetric N-qubit Dicke states	76
	5.2.6	Entanglement at remote distances by using a detection scheme based on optical fibers	77
	5.2.7	Experimental feasibility	77

	5.2.8	Generation of symmetric Dicke states in photon qubits	79
	5.2.9	Conclusions	80
5.3	Generation of arbitrary total angular momentum eigenstates		81
	5.3.1	Introduction: coupling of angular momenta of non-interacting qubits	81
	5.3.2	Description of the physical system	82
	5.3.3	Preparation of total angular momentum eigenstates	83
	5.3.4	Conclusions	88
5.4	Generation of symmetric entangled states by tuning of local operations		90
	5.4.1	Introduction: tripartite entanglement classes of W, GHZ and separable states	90
	5.4.2	Description of the physical system	91
	5.4.3	Generation of 3-qubit W, GHZ and separable states	93
	5.4.4	Generation of N-qubit W, GHZ and separable states	95
	5.4.5	Operational determination of tripartite entanglement classes	96
	5.4.6	Outlook	98

6 Evidence and experimental proof of non-classicality — 99

6.1	Historical introduction to Bell Inequalities		99
6.2	Investigating polarization correlations using CHSH inequalities		101
	6.2.1	Description of the physical system	101
	6.2.2	Derivation of CHSH inequalities for polarization correlations	103
	6.2.3	Violating CHSH inequalities by polarization correlations	104
6.3	Investigating spatial correlations using CHSH inequalities		107
	6.3.1	Derivation of CHSH inequalities for spatial correlations	107
	6.3.2	Violating CHSH inequalities by spatial correlations	108
6.4	Time dependent spatial correlations		110
	6.4.1	Time dependent intensity correlations of second order	110
	6.4.2	Derivation and violation of CHSH inequalities for time dependent spatial correlations	110
	6.4.3	Interpretation	111
6.5	Investigating spatial correlations for multiple emitters ($N > 2$)		114
	6.5.1	Description of the physical system	114
	6.5.2	Intensity correlation signal of second order and its visibility for multiple emitters	115

- 6.5.3 Detection of two photons out of N scattered photons 117
- 6.5.4 Derivation of CHSH inequalities for spatial correlations and multiple emitters $(N > 2)$ 118
- 6.5.5 Violating CHSH inequalities for multiple emitters $(N > 2)$ by spatial correlations 119
- 6.5.6 Conclusion 121
- 6.6 A more suitable inequality for multiple emitters 122
  - 6.6.1 Derivation of a homogeneous Bell-Wigner (HBW) inequality 122
  - 6.6.2 Violation of the HBW inequality for $N \geq 2$ emitters by spatial correlations 123
  - 6.6.3 Conclusion: violation of the HBW inequality and the visibility of the correlated signal 125

7 Conclusions 127

A Expectation values for multi-time intensity correlations 129

B Derivation of Eq. (4.3.2) 131

C Proof of the inequality (6.6.2) 133

Bibliography 135

Chapter 1

Introduction

In this thesis the concept of *entanglement*, although well-known in quantum optics for more than 70 years [1,2], plays a crucial role and is yet about to be implemented in an unconventional way. As it was defined early by Schrödinger, the phenomenon of entanglement is present if the state of a quantum system as described by its wave-function cannot be formulated independently from the state of another system [2]. In an original Gedankenexperiment elaborated by Einstein in 1935, he pointed out correctly that this phenomenon would apply also for *two systems which no longer interact [anymore so that] no real exchange can take place* [1]. The consequences for the state of one system when performing a measurement on the other system he later called *a spooky action at a distance*, displaying his strong disbeliefs in the theory [3].

Today, in historical agreement with Einstein's ideas, the common picture derived when speaking of entanglement still involves two or more particles which may have interacted with each other at some time and space and which thereafter share a joint state. Indeed, for most experiments generating entangled quantum states, previous interactions such as cascade emission [4], non-linear interaction [5], atomic collisions [6,7], Coulomb coupling [8,9], or atom-photon interfaces [10] are a prerequisite. On the other hand, an entangled quantum state can also be prepared solely via a measurement process, where none of the system's constituents must have interacted with each other before [11–20]. The latter method will be investigated throughout this thesis and applied to a simple yet fruitful system giving rise to a large variety of interesting results which we studied recently in a line of publications [19, 21–25].

Let us highlight the basic concept of *measurement-induced entanglement* by using an intuitive approach considering the seminal setup of a Young-type double slit experiment. A basic Young-type double slit setup consists of a double slit aperture being illuminated by coherent light as shown in figure 1.1 [26]. If the dimensions of the slit size and slit separation are chosen appropriately, one can find a characteristic sinusoidal intensity distribution in the far-field region of the illuminated aperture. This diffraction pattern is

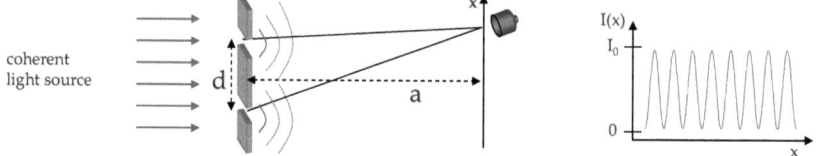

Figure 1.1: Schematic setup of Young's double slit experiment: coherent light shines upon an aperture with two slits separated by a distance d. In the far-field region of the aperture in a distance a, where $a \gg d$, a photon counting device measures the intensity distribution along the x-axis as shown in the picture. Due to interference of the optical waves emanating from the two slits the intensity distribution displays a sinusoidal interference pattern.

commonly explained and described by the interference of electromagnetic waves emanating from the aperture. Indeed, if one slit is blocked and the light thus passes only through the other remaining slit, the interference pattern disappears and only the intensity distribution of a single slit can be seen in the plane of observation. Therefore, the interference observed at a double slit was considered the most important experiment supporting the idea that light must be treated as an electromagnetic wave [26].

However, with Einsteins explanation of the photoelectric effect [27] and the upcoming idea that light consists of indivisible quanta of energy, called photons, the double slit experiment provided an interesting question: will the interference pattern remain when a single-photon emitting light source is used and a single-photon counting device is placed in the far-field region of the aperture? Experimentally, a first answer was given in 1909 [28], when Taylor simulated single photons by using feeble light so that statistically much less than a single photon passed through the aperture on average: as shown in figure 1.2 a.), for low numbers of photons the screen of observation shows a random distribution, while the interference fringes become visible when the number of photons measured is increased. The theoretical answer to this seminal question - which included the paradox how a single photon passing through either of the two double slits could *know* about the presence of the other slit - was given by Paul Dirac in 1927: *each photon interferes only with itself. Interference between two different photons never occurs* [29].

From today's point of view, Dirac's first statement remains still valid. Interestingly, one might notice that it even implied a physical concept which was ahead of its time, namely that of quantum entanglement: invoking the picture of quantum paths the theory of quantum mechanics is capable of describing scenarios where the state of a quantum system must be described by a *superposition* of more than one quantum state. Considering the double slit experiment with a single photon only, we find that this scenario applies, if the detection of the photon is observed in the far-field region. In this case, it is impossible to determine *which way* the photon passed by the aperture using the data acquired by

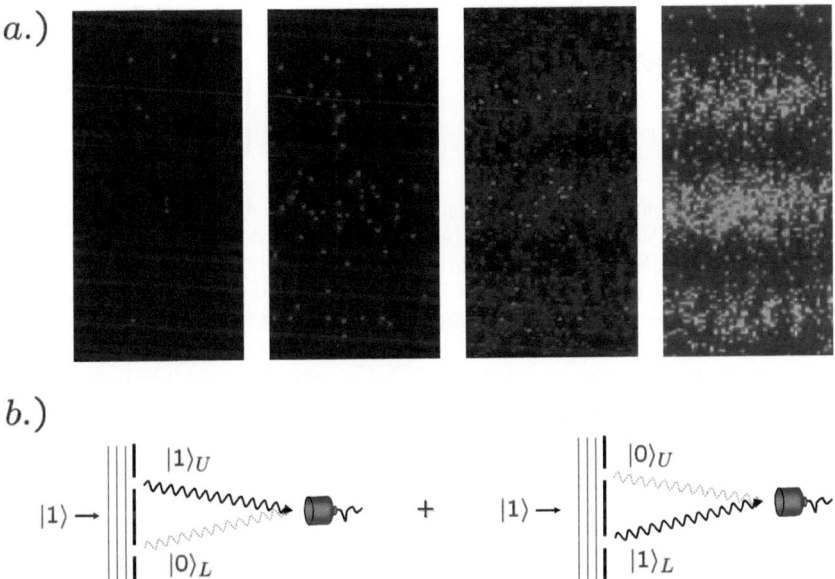

Figure 1.2: *a.*) Measurement data of a CCD-Camera showing the intensity distribution of a double slit experiment with single photons; time of observation is increased from left to right (courtesy of MPQ, Garching).
b.) Theoretical model: quantum paths contributing to the intensity measurement of a double slit experiment. A single photon $|1\rangle$ passing by the aperture can travel through the upper (lower) slit and occupy the mode $|1\rangle_U$ ($|1\rangle_L$) while the lower (upper) mode $|0\rangle_L$ ($|1\rangle_U$) remains unoccupied. Both scenarios displayed are equally probable and provide two possibilities a single photon can pass by the aperture.

the measurement. The scenario is depicted in figure 1.2 *b.*): we denote the path where a photon passes through the upper (lower) slit and none through the lower (upper) slit by the state $|1,0\rangle := |1\rangle_U \otimes |0\rangle_L$ ($|0,1\rangle := |0\rangle_U \otimes |1\rangle_L$). A single photon passing through the double slit aperture is thus described by the superposition $|0,1\rangle + |1,0\rangle$ of the two possible quantum paths which is a not-separable, i.e., an entangled state. Indeed it is possible by using this ansatz to recover the sinusoidal interference pattern of the Young's double slit experiment (see, e.g., [30]). However, in difference to the classical description, it is valid even at the level of single photons.

Throughout this thesis we will encounter similar scenarios, i.e., situations where several distinct quantum paths are possible for a single evolving quantum system. Thereby the intuitive picture of single spatial modes being entangled in their photon occupation numbers will help to describe the correlations occurring between the different quantum paths. We note that the long debate about the actual nature of entanglement for the one-

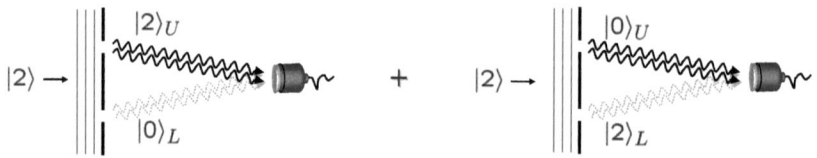

Figure 1.3: Quantum paths contributing to a joint detection measurement of a N00N-state experiment. A bi-photon state $|2\rangle$ can evolve through a double slit aperture by going through the upper (lower) slit and occupying the mode $|2\rangle_U$ ($|2\rangle_L$) while the lower (upper) mode $|0\rangle_L$ ($|1\rangle_U$) remains unoccupied. Hereby, the setup of a N00N-state experiment assures that both photons must travel through the same slit (see, e.g., [33]). Both scenarios displayed are equally probable and provide two possibilities a bi-photon state can pass by the aperture. In an experiment it is required that the detector is sensitive to two-photon events only.

photon entangled state has only recently been settled by considering the field modes and not the photons themselves as being entangled (see [31, 32] and the references therein). Hereby, the concept of single modes being entangled in their occupation numbers can be applied not only to one photon but similarly to two or more photons. This can be used to explain peculiarities which go along with the so-called *N00N*-state, i.e., the occupation of two modes by N photons of the form $|N, 0\rangle + |0, N\rangle$ [33]. For example, a two-mode occupation of a bi-photon state $|2, 0\rangle + |0, 2\rangle$ as depicted in figure 1.3, can result in a sinusoidal pattern in the plane of observation oscillating with a spatial frequency two times as fast as in the case of a two-mode occupation of a one-photon state $|1, 0\rangle + |0, 1\rangle$ [33] (see also section 4.1.1). Obviously, the interference of a bi-photon state is in conflict with Dirac's second statement. However, nowadays plenty of scenarios are known in quantum optics where *interference between two photons does occur* [34–40]. Nevertheless Dirac's second statement remains valid if used in the context of a classical intensity measurement as in Young's double slit experiment.

In this thesis we will investigate a well-defined system of two or more single-photon emitters coherently or incoherently excited and localized in space while looking for correlations within their fluorescence signal. The physical picture used to describe the quantum interferences exhibited by this system is closely linked to the modern concept of a double slit experiment implying single modes being entangled in their occupation numbers. For example, the simplest form of our setup considers $N = 2$ emitters, localized in a trap with a well-defined spacing, which both scatter a single photon. Further, we define two spatial modes by placing two detectors in the far-field of these emitters. Then, if both detectors register jointly an event, we can determine via post-selection that two spatial modes where occupied by two photons. Surprisingly, despite the two scattered photons being initially uncorrelated, this setup turns out to be capable of restoring the results described for the case of a $|2,0\rangle + |0,2\rangle$ photon number occupation and which can even go beyond (see chapter 4).

The thesis is composed of four parts: in chapter 2 and 3, we introduce the basic setup and the theoretical tools needed to describe and characterize the various features displayed by the system. In chapter 4, we then discuss the correlations found in the fluorescence signal of single-photon emitters in the context of *quantum imaging*. There, it will be shown that these correlations can be used to overcome classical signals in terms of resolution when applied to image processing. Understanding the origin of the correlations observed in the fluorescence signal, in chapter 5, we next proceed to a more complex scheme. Introducing multi-level emitters, we utilize the correlations found to project correlated quantum states into the long-lived ground levels of the atoms. Hereby, a broad variety of entangled quantum states can be generated, from the well-known *W-states* and *GHZ-states* to arbitrary total angular momentum eigenstates. Finally, in chapter 6, we end by taking a closer look at the nature of the quantum correlations in the fluorescence light. By looking at various inequalities of classical probability theory, amongst others at Bell and CHSH inequalities, we are able to prove the quantum nature of the correlations.

Chapter 2

Matter-field interaction of single-photon emitters

Our analyses are based on single-photon emitters and offer results which might be applied to any system dealing with such. In this chapter, we introduce the basic setup used for our investigations, its relevant parameters and theoretical tools needed to describe the interplay between single-photon emitters and their scattered photons.

In quantum optics, it is a great goal to find *perfect* single-photon sources, in the sense of an ideal conditionality. However, for our investigations it will be rather irrelevant whether the photon sources themselves are perfect or not, since we always assume a complete measurement cycle, i.e., we count a run as successful if and only if N photons are scattered by N emitters and are registered at N distinct detectors. Any other outcome is rejected and not further regarded. Fast detectors are needed to reduce the observation time of the measurement cycle so that it can be much smaller than the average photon emission time of the light source. In this case, our system can profit from a high repetition rate while it can be realized using a broad range of single-photon sources. Among such sources are trapped ions [8,9,41,42] or neutral atoms [43,44], single molecules [35,45–47], nitrogen-vacancy defect centers in diamond structures [48–50] or quantum dots [34,51,52], where the particular choice depends on the specific application under investigation. However, we believe that the best results are achieved by using trapped ions since their internal and external degrees of freedom can be controlled best at the moment [8,9,41,42]. Therefore, without loss of generality, we will refer to trapped ions when speaking of single-photon emitters in the following.

2.1 Basic system (far-field detection)

The basic system which is used and referred to throughout this thesis involves N atoms ($N \geq 2$) that are localized in space. All N atoms are initially coherently excited by a laser π pulse which transfers all of them to an excited state $|e\rangle$. Subsequently, due to spontaneous decay, all N excited atoms scatter incoherently N single photons. These N photons are registered at N distinct photon detectors placed in the far-field region of the emitters. For the case of $N = 2$ this setup is depicted in figure 2.1.

2.1.1 Description of the measurement cycle

A successful measurement cycle is completed by registering all N scattered photons at the N different photon detectors. Since the N atoms are initially uncorrelated the photons are scattered incoherently and with a random distribution. Our requirement to detect the N photons at N different detectors might thus be described as *probabilistic*. However, the pulse repetition rate of current pulsed lasers can be of the order of several tens of MHz [20], whereas the typical lifetime of trapped atoms is of the order of ten nanoseconds so that the time for one measurement cycle can be of the order of a few tens of nanoseconds. In other words, the system under consideration suffers from a probabilistic behavior, but can profit from a very fast repetition rate. In case that one of the N detectors does not register any photon within the time window of one measurement cycle, which has to be chosen slightly smaller than the repetition rate, that cycle is discarded and a new cycle is started by the next exciting laser π pulse. Restricting ourselves to this measurement cycle bears two advantages: first, it simplifies the calculation and, second, it eases the experimental realization of the setup.

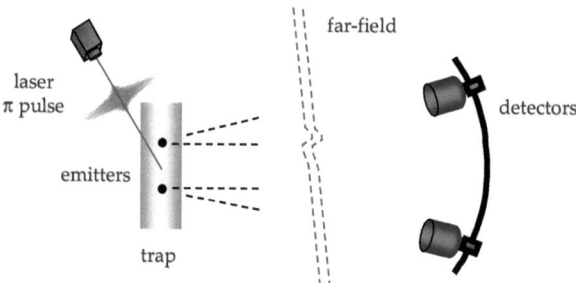

Figure 2.1: Basic setup for $N = 2$ single-photon emitters. Two emitters scatter two photons which are registered at two distinct detectors placed in the far-field region of the emitters.

2.1.2 Quantum paths and loss of which-way information

The detection of all N photons is realized by using N distinct single-photon detectors placed in the far-field region of the emitters. Due to the far-field condition, none of the N detectors can determine which of the N atoms emitted the particular photon it registered. As an example, let us consider the setup for $N = 2$ emitters and detectors as shown in figure 2.1. A successful measurement, where each of the two detectors registers a single photon, can result from the two different yet equally probable quantum paths which are shown in figure 2.2: either the photon scattered by emitter 1 (2) is registered at detector 1 (2) or it is registered at detector 2 (1).

In general, for a setup consisting of N emitters and N detectors placed in the far-field region of the emitters there are $N!$ equally probable and yet different quantum paths, due to permutation of all possible quantum pathways. These quantum paths cannot be distinguished by the information obtained in the measurement cycle and thus all of these $N!$ quantum paths have to be accounted for and contribute to a single successful measurement. This phenomenon is commonly referred to as *loss of which-way information*: if a quantum system can evolve along more than one distinct quantum path and it is furthermore impossible to determine exactly which quantum path was actually chosen using the data obtained in the measurement, then all possible quantum paths weighted with their according probabilities have to be superposed and contribute to the final measurement outcome.

Superpositions of quantum pathways are generally accompanied by quantum interference effects (c.f. chapter 3). The latter can be controlled in the considered system by modifying the setup's geometry. Hence, the system considered here provides us with an important tool that can be fruitfully exploited in different applications. In particular, by manipulating these quantum paths one can engineer overall success probabilities for correlation measurements of more than one photon due to multi-photon interference effects. On the other hand, as will be explained later, the interference between these quantum paths can be used also to engineer particular atomic states of the emitters themselves and to generate specific correlated quantum states.

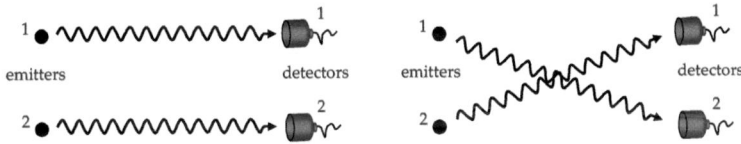

Figure 2.2: Quantum paths leading to a successful measurement for the basic setup implying $N = 2$ emitters and detectors. Both scenarios are equally probable and provide two possibilities the system can evolve along.

2.1.3 Geometry of the setup

The setup introduced so far relies on a far-field detection scheme. This implies already some geometrical considerations which we will characterize in the following, introducing some geometrical measures which will be helpful throughout this thesis.

We denote the position of the nth emitter by $\mathbf{R}_n$ and the position of the jth detector by $\mathbf{r}_j$. Furthermore, all N emitters are considered to be aligned along an axis with equal next-neighbor distance $d := |\mathbf{R}_{n+1} - \mathbf{R}_n|$. As shown in figure 2.3, we choose the origin of our coordinate system to coincide with $\mathbf{R}_0$ which is the virtual extension of this alignment. This allows us to define the position of the nth atom as

$$\mathbf{R}_n = n\,\mathbf{R}_1. \qquad (2.1.1)$$

In analogy to, e.g., the classical description of Young's double slit experiment, we find that the two optical paths connecting two neighboring atoms with one detector placed at $\mathbf{r}_j$ in the far-field region of the emitters differ in an optical phase δ_j which can be described

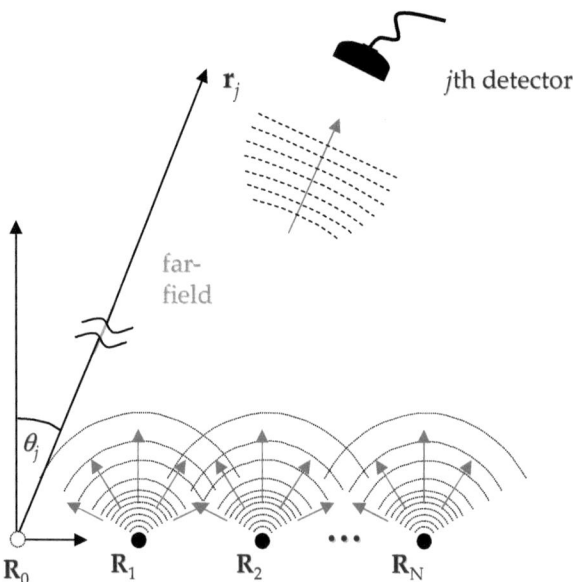

Figure 2.3: Basic setup with far-field requirement showing N emitters and one detector. We choose the origin of the coordinate system at $\mathbf{R}_0$ such that it permits a simple notation of the atoms' positions $\mathbf{R}_n$ (see Eq. (2.1.1)).

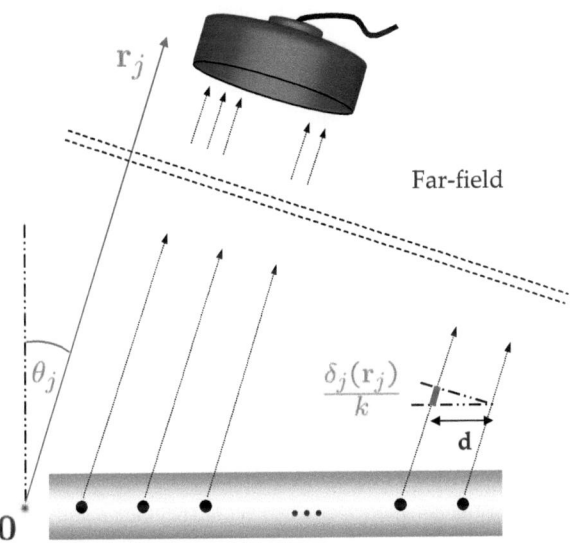

Figure 2.4: Basic setup with far-field requirement showing N emitters and one detector. The coordinate system introduced in figure 2.3 allows for a simple description of the optical phase difference δ_j between the two quantum paths of two neighboring atoms and a detector placed at $\mathbf{r}_j$ in the far-field region, respectively (see also Eq. (2.1.2)).

by

$$\delta_j := \delta(\mathbf{r}_j) = kd \frac{\mathbf{r}_j \cdot \mathbf{R}_1}{|\mathbf{r}_j| |\mathbf{R}_1|} = kd \sin \theta_j, \quad (2.1.2)$$

where k is the wavenumber of the scattered light and θ_j denotes the scattering angle, i.e., the angle between $\mathbf{r}_j$ and a perpendicular to the line of atoms as shown in figure 2.4. As can be seen from the figure, the optical phase $\delta_j = \delta(\mathbf{r}_j)$ depends solely on the position $\mathbf{r}_j$ of the jth detector, if the alignment of the atomic emitters is fixed. Therefore, we are going to interchange $\mathbf{r}_j$ and δ_j in the argument of functions whenever it seems convenient in the remainder of this thesis (see, e.g., Eq. (2.2.15)).

2.1.4 Two-level systems and the atomic projection operator

As pointed out before, it is advantageous to choose trapped atoms as a source for single photons since this source can be controlled experimentally to a great extend. It also allows to describe the physics of our system in the possibly simplest way. For example, one can assume to work with an ideal two-level system, realized by pumping the atoms into an appropriate excited state from which they can decay to a single level only, e.g., to the ground state. Moreover, in case of ions, trapped in an ion trap inside the Lamb-Dicke regime, the recoil effects and other disturbances can be kept small [53, 54].

In the case of a two-level system, we denote the excited state by $|e\rangle$ and the ground state by $|g\rangle$. Furthermore, we assume that the exciting coherent laser transfers all ions along $|g\rangle \to |e\rangle$ via a π pulse. Therefore we can write down the initial state of our N two-level systems, just after the excitation in the pulsed regime, as

$$\prod_{n=1}^{N} |e\rangle_n = |e\rangle_1 \otimes |e\rangle_2 \otimes \ldots \otimes |e\rangle_N. \qquad (2.1.3)$$

This initial state factorizes completely, i.e., it is a separable state, and thus - initially - all emitters can be considered as completely *uncorrelated*.

From the initial state of Eq. (2.1.3), each of the atoms may only decay along the channel $|e\rangle \to |g\rangle$ via spontaneous emission of a photon. However, since we only want to describe those cases where all emitted photons are successfully detected, it is more elegant to describe the process of the decay by the detection of the fluorescent photon. In this way, we link the detection process of a photon with the projection of the atomic emitter state (see, e.g., chapter 7 of [55]).

The link between the detection process and the atomic state projection enables us to introduce a useful tool, namely a detection operator (see section 2.2.5 and Eq. (2.2.14)) which describes the detection of a photon by acting on the initial states $|e\rangle$ of the emitting atoms. More specifically, the atomic projection is described by an operator

$$\hat{S}_n^- := |g\rangle_n \langle e|, \qquad (2.1.4)$$

which projects the nth atomic emitter from its initial state $|e\rangle$ onto its ground state $|g\rangle$. In the following we will consider the evolution of this atomic operator in more detail before we derive a handy expression for the detection operator suitable for our setup in section 2.2.5.

2.1.5 Time evolution of the atomic projection operator

In this section, we examine the dynamics of the setup characterized in section 2.1 which is determined by the dynamics of the atomic projection operators. Hereby, we do not consider any atom-laser interactions since all of our following investigations assume an initial excitation of the atomic emitters in the pulsed regime, i.e., all atoms are transferred initially in the excited state by a laser π pulse whereas after the excitation no laser radiation is present any more while the initial radiation field is described by empty modes of the vacuum.

In general, the process of spontaneous emission of a system of multiple atomic emitters can be treated by solving the corresponding master equation of the density operator of the atomic system. The density operator of any quantum system can be written as

$$\rho := \sum_n p_n |\Psi_n\rangle\langle\Psi_n|, \qquad (2.1.5)$$

where the sum goes over all constituent states $\{|\Psi_n\rangle\}$ of the system and p_n denotes the statistical weight of each state ($\sum_n p_n = 1$). The time evolution of the system can then be derived from the Schrödinger equation $\dot\rho := -\frac{i}{\hbar}[H, \rho]$ with H being the system's total Hamiltonian.

For our system of N two-level atoms we define the following set of operators

$$\hat{S}_n^+ := |e\rangle_n\langle g|, \quad \hat{S}_n^- := |g\rangle_n\langle e|, \quad \hat{S}_n^z := \frac{1}{2}(|e\rangle_n\langle e| - |g\rangle_n\langle g|), \quad n = 1, ..., N. \qquad (2.1.6)$$

As derived, e.g., in [55, 56], by applying Born-, Rotating-Wave- and Markov-approximations, one can derive the master equation for an ensemble of N initially fully excited, identical two-level systems in the Schrödinger picture as

$$\frac{\partial \rho}{\partial t} = -i\omega_0 \sum_n^N \left[\hat{S}_n^z, \rho\right] - \frac{i}{2} \sum_{n \neq m}^N \Omega_{nm} \left[\hat{S}_n^+ \hat{S}_m^-, \rho\right]$$
$$- \sum_{nm}^N \gamma_{nm} \left(\hat{S}_n^+ \hat{S}_m^- \rho - 2\hat{S}_m^- \rho \hat{S}_n^+ + \rho \hat{S}_n^+ \hat{S}_m^-\right), \qquad (2.1.7)$$

where $\omega_0 := \omega + \Omega_{nn}$, $\omega := 2\pi \frac{c}{\lambda}$, $\Omega_{nn} \equiv \Omega_L$ the Lamb shift, λ is the wavelength, Ω_{nm} a level shift due to interatomic dipole-dipole interaction (see section 3.3) and γ_{nm} is related to the decay constant of the atomic transition $|e\rangle \to |g\rangle$ where in particular $\gamma_{nn} = \gamma$ (for further details see, e.g., [55, 56]).

In the following, most of our investigations are restricted to the special case where the interatomic distances $d := |\mathbf{R}_1|$ are large in comparison to the wavelength λ of the transition $|e\rangle \to |g\rangle$. Only in section 3.3 we consider the case where $d \leq \lambda$ which we later apply in section 4.2.3. In case of $d \gg \lambda$, we may approximate $\Omega_{nm} \to 0$ and $\gamma_{nm} \to 0$ for $n \neq m$ so that Eq. (2.1.7) can be simplified to

$$\frac{\partial \rho}{\partial t} = -i\omega \sum_n^N \left[\hat{S}_n^z, \rho\right] - \gamma \sum_n^N \left(\hat{S}_n^+ \hat{S}_n^- \rho - 2\hat{S}_n^- \rho \hat{S}_n^+ + \rho \hat{S}_n^+ \hat{S}_n^-\right). \qquad (2.1.8)$$

From Eq. (2.1.8) we can calculate the equations of motion for the expectation values of the constituents of ρ^1 which we are going to use in the following:

$$\left(\frac{\partial}{\partial t}\right)\langle S_n^\pm\rangle = (\pm i\omega - \gamma)\langle S_n^\pm\rangle, \tag{2.1.9}$$

$$\left(\frac{\partial}{\partial t}\right)\langle\prod_n^m S_n^+ \prod_n^m S_{m-n+1}^-\rangle = -2m\gamma\langle\prod_n^m S_n^+ \prod_n^m S_{m-n+1}^-\rangle, \tag{2.1.10}$$

where $n = 1, ..., m \leq N$. These equation of motion can be easily integrated to

$$\langle S_n^\pm(t)\rangle = e^{(\pm i\omega - \gamma)t}\langle S_n^\pm(0)\rangle, \tag{2.1.11}$$

$$\langle S_n^+(t_1)S_n^-(t_1)\rangle = e^{-2\gamma t}\langle S_n^+(0)S_n^-(0)\rangle, \tag{2.1.12}$$

$$\langle\prod_n^m S_n^+(t) \prod_n^m S_{m-n+1}^-(t)\rangle = e^{-2m\gamma t}\langle\prod_n^m S_n^+(0)S_n^-(0)\rangle. \tag{2.1.13}$$

Furthermore we can make use of the quantum regression theorem (see appendix A) to calculate the expectation values of operators acting at different times as

$$\langle S_n^+(t_2)S_n^-(t_1)\rangle = e^{i\omega(t_2-t_1)-\gamma(t_2+t_1)}\langle S_n^+(0)S_n^-(0)\rangle, \tag{2.1.14}$$

$$\langle\prod_n^m S_n^+(t_n) \prod_n^m S_{m-n+1}^-(t_{m-n+1})\rangle = \prod_n^m e^{-2\gamma t_n}\langle\prod_n^m S_n^+(0) \prod_n^m S_{m-n+1}^-(0)\rangle, \tag{2.1.15}$$

where we assumed the time-ordering $t_m > t_{m-1} > ... > t_1$.

In summary, restricting our setup to interatomic distances of $d \gg \lambda$ simplifies significantly the dynamics of the expectation values of the atomic projection operators $\hat{S}^+$ and $\hat{S}^-$, since the time evolution decouples. This is the case, in particular for the expectation value displayed in Eq. (2.1.15) which we are going to apply in the next section where we investigate intensity correlation functions of Nth order.

---

[1]Note that any constituent of ρ can be expressed by the atomic operators $\hat{S}^+$ and $\hat{S}^-$.

2.2 Correlation functions and detection probabilities

The probabilities to detect all N photons emitted by the system of N two-level atoms introduced in section 2.1 can be calculated by the use of correlation functions of the electric field, generated by the N atomic emitters. In the following, we review the concept of correlation functions in quantum optics as introduced by Glauber in 1963 [57, 58] and their interpretation as detection probabilities. In particular we focus on the derivation of intensity correlation functions of Nth order.

2.2.1 The electric field

Let us turn to the electric field as generated by the chain of N two-level atoms excited in the pulsed regime. The positive frequency part of the electric field amplitude at position $\mathbf{r}_j$ in the far-field region of the emitters can be written in terms of the atomic operators $\hat{S}_n^+(t_j)$ ($\hat{S}_n^-(t_j)$) as [55]

$$\begin{aligned} \mathbf{E}^{(+)}(\mathbf{r}_j, t_j) &= \mathcal{E}_0 \frac{e^{-ik|\mathbf{r}_j|}}{|\mathbf{r}_j|} \sum_{n=1}^{N} e^{ik(\frac{\mathbf{r}_j}{|\mathbf{r}_j|} \cdot \mathbf{R}_n)} \frac{\mathbf{r}_j}{|\mathbf{r}_j|} \times \left(\mathbf{d}_{eg} \times \frac{\mathbf{r}_j}{|\mathbf{r}_j|} \right) \hat{S}_n^- \left(t_j - \frac{|\mathbf{r}_j|}{c} \right) \\ &= \mathcal{E}_0 \frac{e^{-ik|\mathbf{r}_j|}}{|\mathbf{r}_j|} \frac{\mathbf{r}_j}{|\mathbf{r}_j|} \times \left(\mathbf{d}_{eg} \times \frac{\mathbf{r}_j}{|\mathbf{r}_j|} \right) \sum_{n=1}^{N} e^{in\delta_j} \hat{S}_n^- \left(t_j - \frac{|\mathbf{r}_j|}{c} \right), \end{aligned} \qquad (2.2.1)$$

where $\mathcal{E}_0$ and $\mathbf{d}_{eg}$ denote the field's constant amplitude and the dipole moment of the transition $|e\rangle \rightarrow |g\rangle$, respectively. Hereby, we utilized the geometry of the setup introduced in section 2.1 by means of Eq. (2.1.2). The negative frequency part is attainable by complex conjugation, i.e., $\mathbf{E}^{(-)}(\mathbf{r}_j, t_j) = \left(\mathbf{E}^{(+)}(\mathbf{r}_j, t_j)\right)^\dagger$. As before, we neglect any subsequent laser-field interaction so that the initial field is assumed to be in the vacuum state.

2.2.2 Intensity correlation function of first order

Let us look at the expectation values of the amplitude of the electric field[2]. The correlation function of first order $G^{(1)}(\mathbf{r}_1, \mathbf{r}_2; t_1, t_2)$ is defined as the following expectation value [55, 57, 58]

$$G^{(1)}(\mathbf{r}_1, \mathbf{r}_2; t_1, t_2) := \left\langle E^{(-)}(\mathbf{r}_1, t_1) E^{(+)}(\mathbf{r}_2, t_2) \right\rangle, \qquad (2.2.2)$$

---

[2] The electric field of Eq. (2.2.1) is defined as a vector field in order to account for the polarization of the field. However, the basic setup introduced in section 2.1 considers two-level emitters only and thus we neglect the polarization state without loss of generality.

i.e., the *first* order correlation function is of *second* order in the electric field. In this thesis, we will encounter intensity-type correlation measurements only, i.e., measurements where each detector is sensitive exclusively on the intensity defined by

$$\langle I(\mathbf{r},t)\rangle = \frac{\epsilon_0 c}{2} \langle E^{(-)}(\mathbf{r},t)\, E^{(+)}(\mathbf{r},t)\rangle, \qquad (2.2.3)$$

where ϵ_0 is the dielectric constant of the vacuum and c the speed of light. To keep our notations simple and without loss of generality we use natural units in the following, setting $\frac{\epsilon_0 c}{2} \equiv 1$. In addition, we introduce the following concise expression for the intensity correlation function of first order $G^{(1)}(\mathbf{r};t)$ which we are interested in

$$G^{(1)}(\mathbf{r};t) := G^{(1)}(\mathbf{r},\mathbf{r};t,t) = \langle E^{(-)}(\mathbf{r},t)\, E^{(+)}(\mathbf{r},t)\rangle \equiv \langle I(\mathbf{r},t)\rangle. \qquad (2.2.4)$$

Eq. (2.2.4) unveils that the order of the correlation function - as it is being used throughout this thesis - coincides with the order of the intensity function[3].

2.2.3 Intensity correlation functions of higher orders

Having introduced both the first order correlation function of the electric field and the intensity correlation function of first order, we can generalize the concept to correlation functions and intensity correlation functions of higher orders. The correlation function of second order is defined as

$$G^{(2)}(\mathbf{r}_1,\mathbf{r}_2,\mathbf{r}_3,\mathbf{r}_4;t_1,t_2,t_3,t_4) = \\ \langle E^{(-)}(\mathbf{r}_1,t_1)\, E^{(-)}(\mathbf{r}_2,t_2)\, E^{(+)}(\mathbf{r}_3,t_3)\, E^{(+)}(\mathbf{r}_4,t_4)\rangle, \qquad (2.2.5)$$

taking into account spatial as well as temporal correlations. In analogy to Eq. (2.2.4) we define the intensity correlation function of second order

$$G^{(2)}(\mathbf{r}_1,\mathbf{r}_2;t_1,t_2) := G^{(2)}(\mathbf{r}_1,\mathbf{r}_2,\mathbf{r}_2,\mathbf{r}_1;t_1,t_2,t_2,t_1) = \\ \langle E^{(-)}(\mathbf{r}_1,t_1)\, E^{(-)}(\mathbf{r}_2,t_2)\, E^{(+)}(\mathbf{r}_2,t_2)\, E^{(+)}(\mathbf{r}_1,t_1)\rangle, \qquad (2.2.6)$$

which we are going to use extensively in the subsequent investigations.

The generalization of the intensity correlation function to Nth order proceeds analogously.

---

[3] We note that there are different definitions in the literature, too. For example in [59] the order of correlation refers to the order of the electric field.

Accounting for N distinct space-time coordinates it is defined as

$$G^{(N)}(\mathbf{r}_1,...,\mathbf{r}_N;t_1,...,t_N) := G^{(N)}(\mathbf{r}_1,...,\mathbf{r}_N,\mathbf{r}_N,...,\mathbf{r}_1;t_1,...,t_N,t_N,...,t_1) \quad (2.2.7)$$
$$= \left\langle \prod_{j=1}^{N} E^{(-)}(\mathbf{r}_j,t_j) \prod_{j=1}^{N} E^{(+)}(\mathbf{r}_{N-j+1},t_{N-j+1}) \right\rangle.$$

One should note that Eq. (2.2.7) holds for any N and thus includes Eqs. (2.2.4) and (2.2.6). We can analyze the dynamics of our setup in terms of the intensity correlation function employing the electric field amplitude given in Eq. (2.2.1). From this expression we see that the dynamics of our system are solely determined by the time evolution of the atomic projection operators which we calculated explicitly in section 2.1.5. Using Eq. (2.1.15) it is simple to show that the time dependence of the intensity correlation function of Nth order factorizes as

$$G^{(N)}(\mathbf{r}_1,...,\mathbf{r}_N;t_1,...,t_N) = \prod_{j}^{N} e^{-2\gamma\left(t_j - \frac{|\mathbf{r}_j|}{c}\right)} G^{(N)}(\mathbf{r}_1,...,\mathbf{r}_N), \quad (2.2.8)$$

where we made use of the abbreviation $G^{(N)}(\mathbf{r}_1,...,\mathbf{r}_N) := G^{(N)}(\mathbf{r}_1,...,\mathbf{r}_N;0,...,0)$.

In the following chapters of this thesis, we are going to calculate and investigate intensity correlations of arbitrary order. Thereby we are either interested in their spatial variation and behavior or we utilize correlation measurements in order to generate and project long-lived quantum states in the internal degrees of the light emitting atoms. We note that due to the simplifications in Eq. (2.1.8) coming along with $|\mathbf{R}_1| \gg \lambda$, the time dependence of the intensity correlation function of Nth order factorizes as we can see from Eq. (2.2.8). Therefore, without loss of generality but to keep notations simple, we will mention the time dependency in the following only if it is necessary (c.f., e.g., section 3.3 and chapter 6). Otherwise, if it is not specified any further, we refer to Eq. (2.2.8).

2.2.4 Interpretation of intensity correlations for a system of atomic emitters

The intensity correlation function for our basic system (c.f. section 2.1) can be interpreted in an intuitive manner: a successful measurement cycle (c.f. section 2.1.1) requires all N emitted photons to be registered at N different detectors. The intensity correlation function of Nth order thus describes the joint measurement of all N detectors. Therefore, it can be easily connected with the probability $P(\mathbf{r}_1,...,\mathbf{r}_N)$ of detecting the N photons at N different points in space $\mathbf{r}_j$ with $j = 1...N$, which can be written as

$$P(\mathbf{r}_1,...,\mathbf{r}_N) = \frac{C_0^N}{\mathcal{E}_0^{2N}} G^{(N)}(\mathbf{r}_1,...,\mathbf{r}_N). \quad (2.2.9)$$

Here, we also account for the overall success probability $\mathcal{C}_0$ for detecting a single photon scattered spontaneously by an excited two-level atom which is defined as

$$\mathcal{C}_0 := \mu \frac{\Delta\Omega}{4\pi}, \qquad (2.2.10)$$

where μ denotes the quantum efficiency of the detector and $\Delta\Omega$ the solid angle covered by detector surface. Moreover, if the number of emitters M exceeds the order of correlation N we are interested in, the detection probability of Eq. (2.2.9) must be slightly modified. At some points throughout this thesis, it will be convenient to use the normalized intensity correlation function which is defined as

$$g^{(N)}(\mathbf{r}_1, ..., \mathbf{r}_N) := \frac{G^{(N)}(\mathbf{r}_1, ..., \mathbf{r}_N)}{\prod_{j=1}^{N} G^{(1)}(\mathbf{r}_j)}. \qquad (2.2.11)$$

In difference to $G^{(N)}(\mathbf{r}_1, ..., \mathbf{r}_N)$, the normalized intensity correlation function of Nth order can be interpreted as a *relative* probability, i.e., excluding any prefactors corresponding to the overall success probability. For $N = 1$ the normalized intensity correlation function of first order $g^{(1)}(\mathbf{r}_1)$ equals unity. However, in case of $N = 2$, the normalized intensity correlation function of second order can be written as

$$g^{(2)}(\mathbf{r}_1, \mathbf{r}_2) = \frac{P(\mathbf{r}_2|\mathbf{r}_1)}{P(\mathbf{r}_2)}, \qquad (2.2.12)$$

relating the normalized correlation function to the conditional probability $P(\mathbf{r}_2|\mathbf{r}_1) = \frac{P(\mathbf{r}_1, \mathbf{r}_2)}{P(\mathbf{r}_1)}$ of detecting one photon at position $\mathbf{r}_2$, given that another one is detected at position $\mathbf{r}_1$. The latter probability will be used to characterize the non-local character of some results derived in chapter 3.

2.2.5 Correlation functions and the detection operator (two-level system)

Finally, we look in more detail at the link between the detection process of the emitted photon and the atomic state projection already introduced in section 2.1.4. This enables us to establish a useful detection operator for our system. Therefore we reconsider the expression of the positive frequency part of the electric amplitude as given by Eq. (2.2.1): assuming the detection to be polarization sensitive, let $\boldsymbol{\eta}_j$ denote the polarization of the detected signal at the jth detector. The scalar product of the electric field vector of the signal at $\mathbf{r}_j$ and the polarization vector $\boldsymbol{\eta}_j \cdot \mathbf{E}^{(+)}(\mathbf{r}_j)$ can be utilized to define a polarization sensitive detection operator [55, 60]. Introducing a convenient normalization factor of $\sqrt{\frac{1}{N}}$ while omitting the redundant overall phase factor of $e^{-ik|\mathbf{r}_j|}$ which applies to all N atoms equally the detection operator becomes

$$\hat{D}_N(\delta_j) := \frac{1}{\sqrt{N}} \left(\boldsymbol{\eta}_j \cdot \mathbf{E}^{(+)}(\mathbf{r}_j)\right) = \frac{1}{\sqrt{N}} \frac{\mathcal{E}_0}{|\mathbf{r}_j|} \left(\boldsymbol{\eta}_j \cdot \mathbf{d}_{eg}\right) \sum_{n=1}^{N} e^{in\delta_j} \hat{S}_n^-, \qquad (2.2.13)$$

where we made use of Eq. (2.2.1) and the fact that the direction of propagation $\frac{\mathbf{r}_j}{|\mathbf{r}_j|}$ is orthogonal to the plane of the polarization vector $\boldsymbol{\eta}_j$.

For the benefit of a simpler notation, we omit the factor $\frac{1}{|\mathbf{r}_j|}$ in the definition of the detection operator. We note that we assume without loss of generality that $|\mathbf{r}_j|$ = const. for any j due to the far-field requirement in all following considerations. In turn, a factor of $|\mathbf{r}_j|^2$ is omitted in the definition of the overall success probability for our setup (c.f. Eq. (2.2.10)).

The term $(\boldsymbol{\eta}_j \cdot \mathbf{d}_{eg})$ of Eq. (2.2.13) describes the projection of the polarization vector of the electric field onto the polarization vector defined by the polarizer axis of the jth polarizer. We note that a polarization insensitive measurement is also contained in Eq. (2.2.13). This scenario is retained by setting $\boldsymbol{\eta}_j = \mathbf{d}_{eg}$, where all photons are transmitted, so that we obtain

$$\hat{D}_N(\delta_j) := \frac{\mathcal{E}_0}{\sqrt{N}} \sum_{n=1}^{N} e^{in\delta_j} \hat{S}_n^-. \qquad (2.2.14)$$

In chapter 3 and 4, we will always assume the detectors to be insensitive to the signal's polarization and hence we will apply the operator defined in Eq. (2.2.14) which does not account explicitly for the polarization degrees of freedom of the scattered photons.

With Eqs. (2.2.13) and (2.2.14) at hand we can finally formulate the intensity correlation function of Nth order in terms of the detection operator $\hat{D}_N(\delta_j)$

$$G^{(N)}(\mathbf{r}_1, ..., \mathbf{r}_N) = \left\langle \prod_{j=1}^{N} \hat{D}_N^\dagger(\delta_j) \prod_{j=1}^{N} \hat{D}_N(\delta_{N-j+1}) \right\rangle, \qquad (2.2.15)$$

where $\hat{D}_N(\delta_j)^\dagger$ denotes the complex conjugate of $\hat{D}_N(\delta_{N-j+1})$.

Chapter 3

Quantum interferences in the light of single-photon emitters

Some aspects of the setup introduced in chapter 2 have been already investigated earlier in a series of studies [60–63]. For example, the photon statistics revealed by the fluorescence light emitted from a regular chain of atoms was shown to exhibit non-classical behavior [61]. Furthermore, the possibility of observing quantum interferences of second order in the absence of classical interferences (of first order) [60] and its influence on the spatial variation of the spontaneous decay and photon arrival times [62] have been at the focus of former investigations.

With the tools introduced in chapter 2, we are now prepared to investigate quantum interferences of any order in the fluorescence light of the system introduced in section 2.1. We will start by considering the most simple realization of our system consisting of $N=2$ emitters and detectors. This analysis will recall some of the results found in [60–63]. We will then quantify the nature of these quantum interferences by discussing their physical origin and demonstrating their capability of violating Bell-type inequalities.

3.1 Quantum interferences in the fluorescence light of uncorrelated single-photon emitters

For $N=2$ photon emitters, the basic scheme was illustrated in figure 2.1 and is recalled in figure 3.1. Two uncorrelated two-level atoms are localized at positions $\mathbf{R}_1$ and $\mathbf{R}_2$, separated by a distance $d \gg \lambda$ (with λ being the wavelength of the two-level transition). Initially the atoms are excited by a laser π pulse and transferred to the double excited state $|ee\rangle \equiv |e\rangle_1 \otimes |e\rangle_2$ (see also Eq. (2.1.3)). Subsequently, each atom spontaneously emits a single photon which may be detected by two distinct detectors at positions $\mathbf{r}_1$ and $\mathbf{r}_2$. One should note that by restricting the setup to two atoms emitting two photons

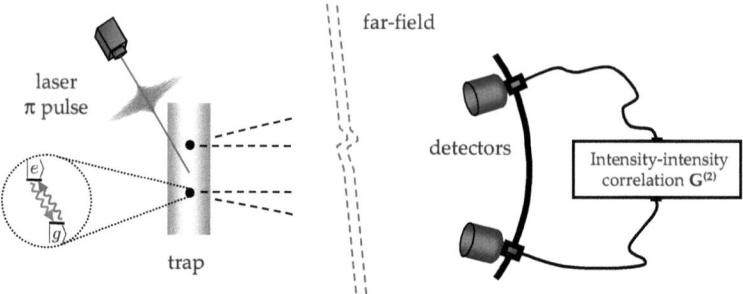

Figure 3.1: Basic setup for $N = 2$ single-photon emitters. Two emitters scatter two photons which are registered at two distinct detectors placed in the far-field region of the emitters.

and demanding that the photons are registered by two distinct detectors, a successful measurement cycle excludes explicitly the possibility of a two-photon absorption at one detector (c.f. section 2.1.1).

Let us first calculate the detection probability $P(\mathbf{r}_1)$ to find a single photon at $\mathbf{r}_1$ which is determined by $G^{(1)}(\mathbf{r}_1)$: with the use of the detection operator defined in Eq. (2.2.14) the intensity correlation function of Eq. (2.2.15) yields

$$\begin{aligned} G^{(1)}(\mathbf{r}_1) &= \langle e,e|\hat{D}_2^\dagger(\delta_1)\hat{D}_2(\delta_1)|e,e\rangle = \left|\hat{D}_2(\delta_1)|e,e\rangle\right|^2 \\ &= \frac{\mathcal{E}_0^2}{2}\left|\left(e^{i\delta_1}|g\rangle_1\langle e| + e^{i2\delta_1}|g\rangle_2\langle e|\right)|e,e\rangle\right|^2 \\ &= \frac{\mathcal{E}_0^2}{2}\left|\left(e^{i\delta_1}|g,e\rangle + e^{i2\delta_1}|e,g\rangle\right)\right|^2 = \mathcal{E}_0^2. \end{aligned} \quad (3.1.1)$$

The detection probability of registering a single photon scattered by either of the two excited atoms is then given by

$$P(\mathbf{r}_1) = \frac{\mathcal{C}_0}{\mathcal{E}_0^2} G^{(1)}(\mathbf{r}_1) = \mathcal{C}_0, \quad (3.1.2)$$

where we made use of Eq. (2.2.9). As a result we find that the detection probability $P(\mathbf{r}_1)$ is equally distributed in space as it exhibits no spatial modulations.

Let us briefly take a look at the first and the last line of Eq. (3.1.1): we calculated $G^{(1)}(\mathbf{r}_1)$ by assuming that both emitters are initially in the excited state. After the detection of one photon at $\mathbf{r}$ one atom is projected onto the ground state while another atom remains in the excited state. However, in a real experiment, measuring $G^{(1)}(\mathbf{r}_1)$ according to Eq. (3.1.1), one thus has to ensure that none of the two atoms has yet scattered a photon due to spontaneous emission. This can be assured either by reading out the atomic state after the measurement and selecting only those results where one of the atoms was found in the excited state or by detecting the second photon which allows similarly to select the proper outcome via post-selection (c.f. section 6.5.3).

Next, we turn to the scenario where two photons emitted by two atoms at $\mathbf{R}_1$ and $\mathbf{R}_2$ are detected by two detectors located at $\mathbf{r}_1$ and $\mathbf{r}_2$. When looking at temporal coincident detection events, the joint detection probability $P(\mathbf{r}_1, \mathbf{r}_2)$ is connected with the intensity correlation function of second order (see Eqs. (2.2.15)) which we calculate to be

$$\begin{aligned} G^{(2)}(\mathbf{r}_1, \mathbf{r}_2) &= \langle ee|\hat{D}_2^\dagger(\delta_1)\,\hat{D}_2^\dagger(\delta_2)\,\hat{D}_2(\delta_2)\,\hat{D}_2(\delta_1)|e,e\rangle = \left|\hat{D}_2(\delta_2)\,\hat{D}_2(\delta_1)|e,e\rangle\right|^2 = \\ &= \frac{\mathcal{E}_0^4}{4}\left|\hat{D}_2(\delta_2)\left(e^{i\delta_1}|g\rangle_1\langle e| + e^{i2\delta_1}|g\rangle_2\langle e|\right)|e,e\rangle\right|^2 \\ &= \frac{\mathcal{E}_0^4}{4}\left|\left(e^{i\delta_2}|g\rangle_1\langle e| + e^{i2\delta_2}|g\rangle_2\langle e|\right)\left(e^{i\delta_1}|g,e\rangle + e^{i2\delta_1}|e,g\rangle\right)\right|^2 \\ &= \frac{\mathcal{E}_0^4}{4}\left|\left(e^{i(\delta_1+2\delta_2)} + e^{i(2\delta_1+\delta_2)}\right)|g,g\rangle\right|^2 \\ &= \frac{\mathcal{E}_0^4}{2}\left(1 + \cos(\delta_2 - \delta_1)\right). \end{aligned} \quad (3.1.3)$$

Moreover, using Eq. (2.2.9) the joint detection probability is thus given by

$$P(\mathbf{r}_1, \mathbf{r}_2) = \frac{\mathcal{C}_0^2}{\mathcal{E}_0^4}\, G^{(2)}(\mathbf{r}_1, \mathbf{r}_2) = \frac{\mathcal{C}_0^2}{2}\left(1 + \cos(\delta_2 - \delta_1)\right). \quad (3.1.4)$$

In result, the intensity correlation function of second order as well as the joint detection probability show a modulation with a contrast[1] of unity depending on the difference in the optical phases δ_1 and δ_2. The intensity correlation function is depicted in figure 3.2.

---

[1] The contrast, i.e., the visibility, of a function $f(x)$ is defined as $[f_{max}(x) - f_{min}(x)]/[f_{max}(x) + f_{min}(x)]$, where f_{max} (f_{min}) is the maximum (minimum) of the function $f(x)$.

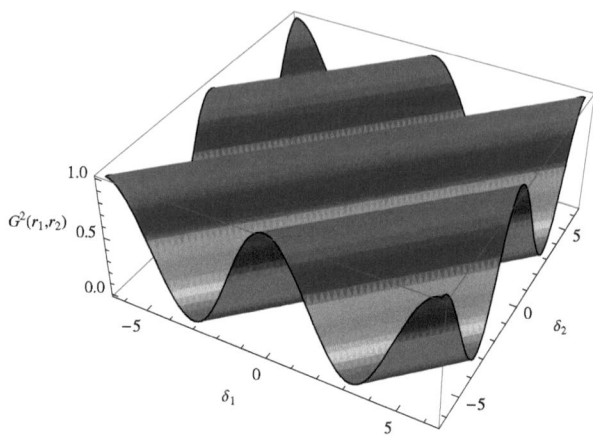

Figure 3.2: Three dimensional Plot (arbitrary units) of the intensity correlation function of second order $G^{(2)}(\mathbf{r}_1, \mathbf{r}_2)$ in dependence of δ_1 and δ_2.

3.2 The physics behind the concept

In the last section, we derived the joint detection probability (second order intensity correlation) for the system introduced in section 2.1. Here, we want to identify the underlying physics behind the calculated result and eventually explain the outcome in terms of quantum interferences.

Let us recall the main result of the foregoing section by considering the conditional probability $P(\mathbf{r}_2|\mathbf{r}_1)$ for the system depicted in figure 3.1. In the case where the two detectors are located at $\mathbf{r}_1$ and $\mathbf{r}_2$ we obtain

$$P(\mathbf{r}_2|\mathbf{r}_1) = \frac{P(\mathbf{r}_1, \mathbf{r}_2)}{P(\mathbf{r}_1)} = \frac{\mathcal{C}_0}{2} \left(1 + \cos\left[\delta_2 - \delta_1\right]\right), \qquad (3.2.1)$$

where we made use of Eqs. (3.1.1) and (3.1.3).

Since Eq. (3.2.1) displays a pure modulation of a cosine with a visibility of 100%, it implies that the conditional probability to find a single photon at position $\mathbf{r}_2$ depends strongly on where the first photon is detected at $\mathbf{r}_1$: this result strikingly violates the concept of locality, i.e., the idea that a measurement at $\mathbf{r}_2$ does not depend on a measurement at $\mathbf{r}_1$ for $\mathbf{r}_2 \neq \mathbf{r}_1$, as it includes, e.g., the case where this probability can be zero. In other words, given that a photon is detected at $\mathbf{r}_1$, then the probability to find another photon at $\mathbf{r}_2$ for $\delta_2 = \delta_1 + (2m+1)\pi$ is *zero*, while for $\delta'_2 = \delta_1 + n\, 2\pi$ it is very likely to find one at $\mathbf{r}'_2$ (m, n being integers).

The reason for this highly non-local behavior, for which no classical analog exists, was

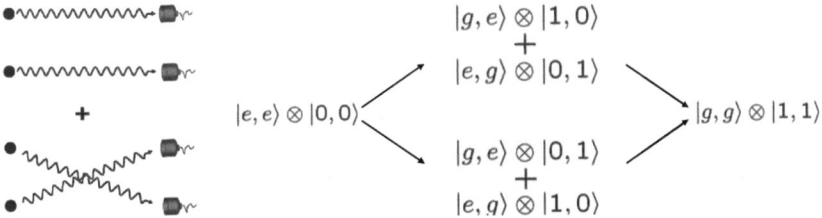

Figure 3.3: Quantum paths which lead to a successful measurement for the basic setup of section 2.1 with $N = 2$ atomic emitters. The two paths connect the initial state $|e,e\rangle \otimes |0,0\rangle$ with the final state $|g,g\rangle \otimes |1,1\rangle$.

already addressed in figure 2.2 and is recalled in figure 3.3. The two detection events take place at two different positions in the far-field region of the emitters. Due to the far-field requirement, none of the two detectors is able to resolve which of the two atoms emitted the particular photon it registered. A successful measurement, where both detectors each register a single photon, can thus result from two different yet equally probable quantum paths shown in figure 3.3: either the photon scattered by emitter 1 (2) is registered by detector 1 (2) or it is registered by detector 2 (1). Thereby, the two photon events cannot be considered independently. One rather has to use a picture involving two-photon amplitudes, i.e., two photon *quantum paths*.

Let us introduce a handy notation to identify the different quantum paths which contribute to a successful measurement: obviously any global state of our system contains information about the state of the atoms, i.e., $|xy\rangle := |x\rangle_1 \otimes |y\rangle_2$ ($x, y \in \{e, g\}$), where the subscripts refer to two atoms located at $\mathbf{R}_1$ and $\mathbf{R}_2$, respectively[2]. In addition, it is coupled with information about the detection events, i.e., $|mn\rangle := |m\rangle_1 \otimes |n\rangle_2$ ($m, n \in \{0, 1\}$), where, e.g., $|1,0\rangle$ ($|0,1\rangle$) describes a state where one photon was scattered and is going to be registered at a detector located at $\mathbf{r}_1$ ($\mathbf{r}_2$) and none at $\mathbf{r}_2$ ($\mathbf{r}_1$). Hence, we denote the initial state where both atoms are in the excited state and none of the detectors has registered a photon so far by $|e,e\rangle \otimes |0,0\rangle$. Likewise, the final state is denoted by $|g,g\rangle \otimes |1,1\rangle$, which accounts for the fact that both detectors registered a single photon while the emitting atoms are projected into their ground states. In figure 3.3 we illustrate the two distinct quantum paths which mediate between the initial and the final state, using the notation introduced above:

1. quantum path: $|e,e\rangle \otimes |0,0\rangle \rightarrow (|g,e\rangle \otimes |1,0\rangle + |e,g\rangle \otimes |0,1\rangle) \rightarrow |g,g\rangle \otimes |1,1\rangle$

2. quantum path: $|e,e\rangle \otimes |0,0\rangle \rightarrow (|g,e\rangle \otimes |0,1\rangle + |e,g\rangle \otimes |1,0\rangle) \rightarrow |g,g\rangle \otimes |1,1\rangle$

---

[2]We note that coherent superpositions of the atomic states $|e\rangle$ and $|g\rangle$ are projected out by the final joint detection measurement of both photons and are thus not further considered.

Hereby, both paths are equally probable. However, depending on the relative position of the two detectors, the bi-photon state may accumulate a different optical phase. In case of the first (second) quantum path, one photon is emitted from an atom at $\mathbf{R}_1$ and detected at $\mathbf{r}_1$ ($\mathbf{r}_2$) accumulating a total optical phase of Φ_{11} (Φ_{12}) and a second photon is emitted from an atom at $\mathbf{R}_2$ and detected at $\mathbf{r}_2$ ($\mathbf{r}_1$) accumulating a total optical phase of Φ_{22} (Φ_{21}). The relative optical phase δ_{12} between the two quantum paths is thus given by

$$\begin{aligned} \delta_{12} &= (\Phi_{11} + \Phi_{22}) - (\Phi_{12} + \Phi_{21}) = \\ &= (\Phi_{11} + \Phi_{12} + \delta_2) - (\Phi_{12} + \Phi_{11} + \delta_1) = \delta_2 - \delta_1, \end{aligned} \quad (3.2.2)$$

where we made use of the definition of δ_i as given in Eq. (2.1.2).

Recalling the intensity correlation function of second order as derived in Eq. (3.1.3), we now see that the modulation observed therein can be associated with the *interference* of two two-photon quantum pathways with a relative optical phase δ_{12}. In this respect it is appropriate to speak of a *quantum interference* phenomenon, in particular of quantum interference of a bi-photon state. By changing the geometry of the setup and especially by varying the detector positions $\mathbf{r}_1$ and $\mathbf{r}_2$ (c.f. Eq. (2.1.2)), the interference between the two quantum paths can thus be observed as exhibited by Eq. (3.1.3).

3.3 Quantum interferences in the regime of dipole-dipole interaction

In section 2.1.5 we investigated the time evolution of the expectation values of the atomic projection operators. We found that the time dependency factorizes in the limit of $d \gg \lambda$, i.e., if the dipole-dipole interaction is negligible. Here, in analogy to section 3.1, we want to derive the normalized intensity correlation function of second order for the more general case where we include the regime where $d \leq \lambda$. In this case, we have to account explicitly for the dipole-dipole term in our derivation and extend the matter-field interactions of our scheme accordingly.

Again, we consider two identical two-level atoms at fixed positions $\mathbf{R}_1$ and $\mathbf{R}_2$. From the seminal work of Dicke [64] we know that dipole-dipole correlations between the atoms are produced if the interatomic spacing $d = |\mathbf{R}_2 - \mathbf{R}_1|$ is of the order of or below the transition wavelength λ (see also section 2.1.5). In this case, the atomic system can be described conveniently in the Dicke basis $\{|e\rangle := |e,e\rangle, |g\rangle := |g,g\rangle, |s\rangle := \frac{1}{\sqrt{2}}(|e,g\rangle + |g,e\rangle), |a\rangle := \frac{1}{\sqrt{2}}(|e,g\rangle - |g,e\rangle)\}$, which forms a set of eigenstates of the combined atomic system. The dipole-dipole interaction between the atoms leads to a level-shift Ω_{12} between the two eigenstates $|s\rangle$ and $|a\rangle$ modifying the unperturbed energies of $|s\rangle$ and $|a\rangle$ accordingly.

The additional correlations between the atomic dipoles are generated because the emission or absorption of a photon cannot be assigned to a specific emitter any longer as a unique labeling of the atoms becomes impossible in this region ($d \leq \lambda$). The atoms are thus indistinguishable and the atom-field interaction has to be symmetrized with respect to atomic permutation. As a consequence, an imbalance appears among the possible transitions between the Dicke states that causes the atomic system to decay with different rates, one enhanced and the other one reduced with respect to the free-atom spontaneous emission rate [55, 56, 62, 64]. The decay rates towards and from the symmetric state $|s\rangle$ and the antisymmetric state $|a\rangle$ differ by $\gamma(|s\rangle) - \gamma(|a\rangle) = 2\Delta\gamma$, where $\Delta\gamma$ is given by

$$\Delta\gamma = \frac{3\gamma}{2(dk)^2}\left(\cos(dk) + \sin(dk)\left(\frac{1}{dk} - dk\right)\right). \quad (3.3.1)$$

In order to derive the dynamics of the expectation values of the atomic projection operators, we have to reconsider the master equation of our setup introduced in Eq. (2.1.7). For two atoms it reads

$$\frac{\partial \rho}{\partial t} = -i\omega \sum_{n=1}^{2}\left[\hat{S}_n^z, \rho\right] - i\frac{\Omega_{12}}{2}\sum_{n\neq m}^{2}\left[\hat{S}_n^+ \hat{S}_m^-, \rho\right]$$
$$- \sum_{nm}^{2} \gamma_{nm}\left(\hat{S}_n^+ \hat{S}_m^- \rho - 2\hat{S}_m^- \rho \hat{S}_n^+ + \rho \hat{S}_n^+ \hat{S}_m^-\right), \quad (3.3.2)$$

where now $\Omega_{12} \neq 0$ and $\gamma_{nm} = \Delta\gamma$ for $n \neq m$ ($n, m = 1, 2$). From Eq. (3.3.2) we can derive the spatial and temporal dependences of the intensity correlation function of first order [65]

$$G^{(1)}(\mathbf{r}, t) = \left\langle \hat{D}_2^\dagger(\mathbf{r}, t) \hat{D}_2(\mathbf{r}, t) \right\rangle = \qquad (3.3.3)$$
$$= \mathcal{E}_0^2 e^{-2(\gamma - \Delta\gamma)t} \left(1 - \frac{1}{2} \frac{\gamma + \Delta\gamma}{\gamma - \Delta\gamma} \left(1 - e^{-4\Delta\gamma t} \right) + \frac{2\Delta\gamma^2}{\gamma^2 - \Delta\gamma^2} \left(1 - e^{-2(\gamma + \Delta\gamma)t} \right) \right).$$

Solving the equations of motion for the normalized intensity correlation function of second order reveals a more compact expression than the unnormalized correlation function[3] as we obtain [62]

$$g^{(2)}(\mathbf{r}_1, 0; \mathbf{r}_2, \tau) = \frac{G^{(2)}(\mathbf{r}_1, 0; \mathbf{r}_2, \tau)}{G^{(1)}(\mathbf{r}_1, 0)\, G^{(1)}(\mathbf{r}_2, \tau)} = \qquad (3.3.4)$$
$$e^{-2\gamma\tau} \left(e^{-2\Delta\gamma\tau} \cos^2\frac{\delta_1}{2} \cos^2\frac{\delta_2}{2} + e^{2\Delta\gamma\tau} \sin^2\frac{\delta_1}{2} \sin^2\frac{\delta_2}{2} + \sin\delta_1 \sin\delta_2 \frac{\cos(\Omega_{12}\tau)}{2} \right).$$

From this expression we can see that the time dependence of the intensity correlation function of second order does not factorize in the limit of $d \leq \lambda$. Nevertheless, in the limit of an almost coincident detection of both photons, i.e., if $\tau \to 0$, the intensity correlation function can be simplified to

$$\lim_{\tau \to 0}(g^{(2)}(\mathbf{r}_1, 0; \mathbf{r}_2, \tau)) \approx \cos^2\frac{\delta_1}{2} \cos^2\frac{\delta_2}{2} + \sin^2\frac{\delta_1}{2} \sin^2\frac{\delta_2}{2} + \frac{1}{2} \sin\delta_1 \sin\delta_2,$$
$$= \frac{1}{2}(1 + \cos(\delta_2 - \delta_1)) = g^{(2)}(\mathbf{r}_1, \mathbf{r}_2), \qquad (3.3.5)$$

which yields essentially the same result as for the intensity correlation function of second order in the case where $d \gg \lambda$ (c.f. Eq. (3.1.3)). There is a simple interpretation for this outcome: if we limit a successful measurement only to almost coincident detection times of both photons, i.e., $\tau \to 0$, then the dipole-dipole interaction cannot affect the intermediate state of the two atoms. Therefore, one might expect a similar outcome for the general case of N emitters and N detectors, if limited to $\tau_1 \to 0, ..., \tau_{N-1} \to 0$, i.e., $\lim_{\tau_1 \to 0, ..., \tau_{N-1} \to 0} (g^{(N)}(\mathbf{r}_1, 0; ...; \mathbf{r}_N, \tau_{N-1})) = g^{(N)}(\mathbf{r}_1, ..., \mathbf{r}_N)$.

This result is applied in section 4.2 where we utilize the spatial dependency of the intensity correlation function for an imaging technique: there, we show that the resolution of an ordinary far-field imaging setup can be exceeded by a factor of N, when imaging atoms with an interatomic spacing of $d \gg \lambda$, by utilizing intensity correlation functions of Nth order. If the interatomic spacing is of the order or smaller than the wavelength, i.e., $d \leq \lambda$, we can then retain this result by restricting the time window of the measurement

---

[3]We note that $G^{(2)}(\mathbf{r}_1, 0; \mathbf{r}_2, \tau)$ displays the same spatial modulations as $g^{(2)}(\mathbf{r}_1, 0; \mathbf{r}_2, \tau)$ since $G^{(1)}(\mathbf{r}, t)$ does not exhibit any spatial modulations (c.f. Eq. (3.3.3)).

cycle so that the approximation of Eq. (3.3.5) is justified.

3.4 A proof of quantum correlations using Bell inequalities

In the present chapter, so far we have derived the intensity correlation function of second order for the basic setup introduced in section 2.1. The interpretation of our results in section 3.2 unveiled that the spatial modulation of the correlation signal arises from the interference of two quantum pathways the whole system can evolve along and which thus have to be superposed to describe the measurement outcome. In this section, we prove the quantum nature of these correlations, i.e., we show that they are due to correlations between the emitted photons which are non-local in character and form an *entangled* state, by demonstrating that Bell-type inequalities can be violated potentially.

An historical introduction to Bell inequalities and a more detailed analysis of the violation will be provided in chapter 6. Here, we focus on our basic system as introduced in section 2.1. In order to apply the Bell inequalities we have to determine the single and joint detection probabilities of our system. These can be calculated by using the intensity correlation function of first or second order, respectively, as derived in chapter 2. We note that the normalization of the correlation function did not play a major role in the investigations carried out so far. However, if we want to apply Bell inequalities, we have to account for an exact normalization.

3.4.1 Derivation of Bell inequalities for single-photon emitters

In his seminal paper Bell proved that deterministic local theories with hidden variables are incompatible with quantum mechanics [66]. Here, we briefly follow his reasoning and start considering the assumptions of a deterministic hidden variable theory. The continuous set of hidden variables is denoted by λ. The probability of registering one photon at the jth detector located at position $\mathbf{r}_j$ is then determined by $p(\mathbf{r}_j, \lambda)$ ($j = 1, 2$), where we included the hidden variables λ in the argument of the single detection probability $p(\mathbf{r}_j)$. Following the requirement of *locality*, the joint probability $p_{12}(\mathbf{r}_1, \mathbf{r}_2, \lambda)$ of detecting two photons at detectors 1 and 2 which are located at $\mathbf{r}_1$ and $\mathbf{r}_2$, respectively, can be written as the product of the two *independent* single detection probabilities

$$p_{12}(\mathbf{r}_1, \mathbf{r}_2, \lambda) = p(\mathbf{r}_1, \lambda) \cdot p(\mathbf{r}_2, \lambda). \quad (3.4.1)$$

Though λ are hidden variables of a deterministic local theory and thus unknown, the detection probabilities obtained when performing a real experiment are determined by the ensemble averages over all λ

$$p(\mathbf{r}_j) = \int d\lambda\, g(\lambda)\, p(\mathbf{r}_j, \lambda) \quad \text{with} \quad j = 1, 2,$$
$$p_{12}(\mathbf{r}_1, \mathbf{r}_2) = \int d\lambda\, g(\lambda)\, p(\mathbf{r}_1, \lambda)\, p(\mathbf{r}_2, \lambda), \qquad (3.4.2)$$

where $g(\lambda)$ denotes a weight function of the hidden variables.

At this point, we make use of the following mathematical inequalities as introduced by Clauser and Horne [67],

$$-XY \leq xy - xy' + x'y + x'y' - Yx' - Xy \leq 0, \qquad (3.4.3)$$

which hold for any values of x, x', y, y', X, Y fulfilling the restrictions $0 \leq x, x' \leq X$ and $0 \leq y, y' \leq Y$. By setting $X = Y = 1$ and identifying

$$p(\mathbf{r}_1, \lambda) = x, \quad p(\mathbf{r}_1', \lambda) = x',$$
$$p(\mathbf{r}_2, \lambda) = y, \quad p(\mathbf{r}_2', \lambda) = y', \qquad (3.4.4)$$

the inequalities of Eq. (3.4.3) read, after multiplying with $g(\lambda)$ and integrating over λ

$$-1 \leq p_{12}(\mathbf{r}_1, \mathbf{r}_2) - p_{12}(\mathbf{r}_1, \mathbf{r}_2') + p_{12}(\mathbf{r}_1', \mathbf{r}_2)$$
$$+ p_{12}(\mathbf{r}_1', \mathbf{r}_2') - p(\mathbf{r}_1') - p(\mathbf{r}_2) \leq 0. \qquad (3.4.5)$$

Eqs. (3.4.5) constitute Bell inequalities with respect to position variables. We are going to use them in order to analyze the correlations between the emitted photons within the system depicted in figure 3.1. Please note that the inequalities (3.4.5) contain four parameters $\mathbf{r}_1, \mathbf{r}_1', \mathbf{r}_2, \mathbf{r}_2'$, i.e., if one is interested to test the inequalities in an experiment, one has to account for at least four different detector positions at which the joint and single probabilities have to be determined independently.

We are going to use the detection probabilities calculated in section 3.1 to test the Bell-type inequalities for our system. The usual conclusion drawn from a violation of such inequalities is that the measured joint detection signal of the system exhibits correlations which cannot be explained by a local deterministic theory with hidden variables and thus may be considered as a proof that quantum mechanics is the correct theory describing reality. Let us moreover emphasize that these correlations cannot be explained in any classical way. However, accepting the theory of quantum mechanics, we may conclude that they exhibit an *entangled* nature.

3.4.2 Violating Bell inequalities by position correlations

For the setup described in section 2.1, we derived the detection probabilities to find a single photon or two jointly detected photons in section 3.2: the chance of finding a single photon at the jth detector located at position $\mathbf{r}_j$, depends on the solid angle $\Delta\Omega$ covered by the detector's surface and its quantum efficiency μ. As derived in Eq. (3.1.2), the single detection probability $p(\mathbf{r}_j)$ is given by

$$p(\mathbf{r}_j) \equiv P(\mathbf{r}_j) = \frac{\mathcal{C}_0}{\mathcal{E}_0^2} G^{(1)}(\mathbf{r}_j) = \mu \frac{\Delta\Omega}{4\pi}. \tag{3.4.6}$$

Whereas, using Eqs. (2.2.9) and (3.1.3), the joint probability $p_{12}(\mathbf{r}_1, \mathbf{r}_2)$ of detecting two photons at $\mathbf{r}_1$ and $\mathbf{r}_2$ is determined as

$$p_{12}(\mathbf{r}_1, \mathbf{r}_2) \equiv P(\mathbf{r}_1, \mathbf{r}_2) = \frac{\mathcal{C}_0^2}{2} \left(1 + \cos(\delta_2 - \delta_1)\right), \tag{3.4.7}$$

By substituting Eqs. (3.4.6) and (3.4.7) into the inequalities of Eq. (3.4.5) and by choosing one of the following two sets of parameters [59, 67, 68]

$$a.) = \begin{cases} \delta_2 - \delta_1 = \frac{3}{8} 2\pi, & \delta_2' - \delta_1 = \frac{1}{8} 2\pi, \\ \delta_2 - \delta_1' = \frac{3}{8} 2\pi, & \delta_2' - \delta_1' = \frac{3}{8} 2\pi, \end{cases} \tag{3.4.8}$$

$$b.) = \begin{cases} \delta_2 - \delta_1 = \frac{1}{8} 2\pi, & \delta_2' - \delta_1 = \frac{3}{8} 2\pi, \\ \delta_2 - \delta_1' = \frac{1}{8} 2\pi, & \delta_2' - \delta_1' = \frac{1}{8} 2\pi, \end{cases} \tag{3.4.9}$$

we obtain the following Bell inequalities

$$-1 \leq (\mathcal{C}_0^2 (\pm\sqrt{2} + 1) - 2\mathcal{C}_0) \leq 0, \tag{3.4.10}$$

where the plus (minus) sign holds for the set of parameters b. (a.).
Both inequalities in (3.4.10), the lower and the upper one, are difficult to violate in an experiment: the upper inequality can be reduced to

$$\mathcal{C}_0 (\pm\sqrt{2} + 1) - 2 \leq 0, \tag{3.4.11}$$

which (assuming $\mathcal{C}_0 \geq 0$) can be violated if

$$\mathcal{C}_0 > \frac{2}{\sqrt{2} + 1} \approx 0.83, \tag{3.4.12}$$

while the lower inequality is violated for

$$\mathcal{C}_0 > \frac{1-\sqrt{\sqrt{2}}}{1-\sqrt{2}} \approx 0.46 \ . \tag{3.4.13}$$

Both cases are impractical to achieve in a real experiment, since in the ansatz used throughout this thesis the overall success probability $\mathcal{C}_0$ for a single detection event is extremely small. In the applications discussed in chapter 4 or 5, the low success probability $\mathcal{C}_0$ can be compensated by the fast repetition rate of the experiment. However, for violating the Bell inequalities (3.4.10), the repetition rate is rather irrelevant: these inequalities can be violated only if either (3.4.12) or (3.4.13) is fulfilled, which both depend on the overall success probability $\mathcal{C}_0$ for detecting a single photon in the far-field of the two possible emitters.

Nevertheless, Eqs. (3.4.12) and (3.4.13) show that it is *potentially* possible to violate the Bell inequalities which - in theory - is sufficient to testify the entangled nature of the photons described before. In chapter 6 we will consider more elaborate Bell inequalities which will overcome these experimental insufficiencies just described and as such can be used demonstrate the entangled nature of the emitted photons also in a real experiment.

3.5 Alternative detection scheme based on optical fibers

In the last section of the present chapter we briefly propose an alternative setup, comparable to the one introduced in section 2.1, but overcoming the constraint of photon detection in the far-field region of the emitters. In particular, in chapter 5 we will encounter situations where the far-field requirement can be considered as a handicap. There we will focus on engineering entangled quantum states within the ground states of the atomic emitters. In this case it will be seen as an advantage if the system can be modified such that it allows also for the entanglement of *remote* atoms, i.e., atoms which may be separated by macroscopic distances.

The far-field requirement can be avoided by the use of optical fibers. Hereby, the atomic emitters may be isolated from each other, but the optical link between the emitters and the detectors is realized by optical fibers. The modified setup is illustrated in figure 3.4 for the case of two emitters and two detectors: one end of a fiber is adjusted such as to collect the photons emitted from one of the atoms, the other end is fed to a bucket detector. Each detector receives two separate fibers linked to both emitters so that again a detection event cannot unveil along which of the two possible paths the registered photon was traveling.

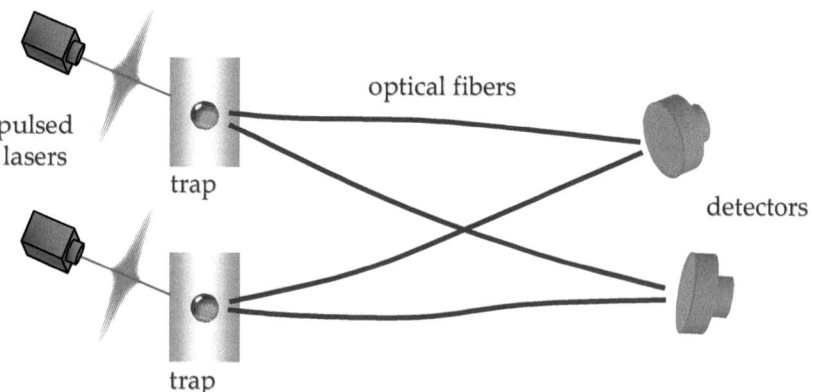

Figure 3.4: Experimental setup for coupling of two atoms via projective measurements using optical fibers. In a successful measurement cycle, each atom emits a single photon and each detector registers a single photon. However, the detectors cannot distinguish which of the atoms emitted the registered photons and thus a correlated measurement can exhibit quantum interferences.

The scheme works analogously for N emitters and N detectors by using N^2 optical fibers to establish all $N!$ quantum paths, along which the system can evolve. Coupling single photons into fibers is experimentally feasible and has been already applied in related experiments (see e.g. [20, 38, 44, 69]). We note that the geometry of the system, i.e., the positions of the emitters and detectors, becomes irrelevant for the measurement outcome if using optical fibers. The optical path differences can be adjusted by manipulating the optical fibers themselves, e.g., by implementing a phase shift within the fiber or at the input. We will see in chapter 5 that by the use of optical fibers new degrees of freedom can be accessed which can be fruitfully exploited in the context of state engineering.

Chapter 4

Quantum imaging using single-photon emitters

In modern quantum optics, there is a great variety of proposals aiming at an improvement of different aspects of the image formation process, commonly summarized by the field of *quantum imaging*. Today, this fast growing field ranges from early Ghost imaging [70], sub-wavelength phase measurements [71,72] to quantum lithography [33,73,74], quantum microscopy [19,75–77] and many more (see, e.g., [78,79]). Though all of these proposals commonly aim to overcome the classical boundaries of the image formation process, only few improve the spatial resolution itself, i.e., the ability to image a physical object while overcoming the Rayleigh [80] or Abbe limit [81] of classical optics. The latter will be at the focus of the following investigations adopting the system described before to implement an image processing technique with focus on microscopic applications.

4.1 Introduction to quantum imaging and Abbe's criterion of resolution

In Young's double slit experiment the probability $P(\mathbf{r}) \propto I(\mathbf{r})$ to detect a photon at position $\mathbf{r}$ in the far-field of the two slits results from the interference of the two possible pathways a single photon can propagate along to reach the detector. Hence, a photon passing through a double slit aperture can be expressed by the state

$$|\psi(1)\rangle = \frac{1}{\sqrt{2}} \left(|1\rangle_U |0\rangle_L + |0\rangle_U |1\rangle_L \right), \quad (4.1.1)$$

where the subscript L (U) denotes the path through the lower (upper) slit (c.f. figure 1.2). A variation of the detector position $\mathbf{r}$ leads to a spatially modulated intensity pattern

$$I(\mathbf{r}) = \frac{I_0}{2}\left(1 + \cos\delta(\mathbf{r})\right), \qquad (4.1.2)$$

where $\delta(\mathbf{r}) = kd\sin\theta(\mathbf{r})$ is the optical phase difference of the waves emanating from the two slits and $k = 2\pi/\lambda$, d, $\theta(\mathbf{r})$ and I_0 denote the wavenumber, slit separation, scattering angle and the constant intensity at $\sin\theta(\mathbf{r}) = 0$, respectively.

Young's double slit experiment represents a far-field imaging setup, where the fringe spacing of the modulation (for fixed d) is determined by the optical wavelength λ. Eq. (4.1.2) allows us to recover Abbe's criterion for resolving an aperture from its intensity distribution $I(\mathbf{r})$ in the far-field, i.e., in the Fourier plane: the aperture can be imaged only if it is possible to measure the intensity pattern in the range $-2\pi \leq kd\sin\theta(\mathbf{r}) \leq 2\pi$, i.e., if we are able to obtain the first principal maxima of the diffraction pattern in the far-field. In the words of Abbe: *we can reconstruct an image from an object if and only if the first diffraction order in the Fourier plane is at least visible* [81]. Otherwise, if we try to image a double slit (or an N-slit) aperture with a smaller spacing, i.e., in the parameter region where the criterion is violated so that the diffraction pattern in the Fourier plane does not display the first principal maxima, the image will start to blur.

Note that Abbe's criterion allows a definition of the resolution limit in any imaging technique (classical or non-classical). In particular, this criterion enables us to compare the classical limit obtained by measuring $I(\mathbf{r})$ with the resolution achieved with specific two-photon correlation signals as recently proposed in the field of quantum imaging: there it is known that quantum interferences based on entangled photon number states are able to overcome the classical resolution limit obtained by measuring the intensity distribution [33, 74, 76, 82–86]. In the following we will describe two representative systems that utilize entangled photons for their image processing purposes.

4.1.1 Example 1: N00N-state lithography

The basic concept of classical lithography can be explained in a neat, simplified toy model [87]: as discussed above, by the use of a double slit aperture one can generate intensity patterns with a single sinusoidal modulation. Using a substrate with an appropriate single-photon absorbing medium, this pattern can be imprinted onto the substrate. Furthermore, an arbitrary pattern can be imprinted onto the substrate by decomposing the desired signal into its Fourier components. The needed Fourier components can be encoded individually on the substrate utilizing appropriate double slit apertures. The desired structure is obtained thereafter by superposing all individual patterns onto the single substrate.

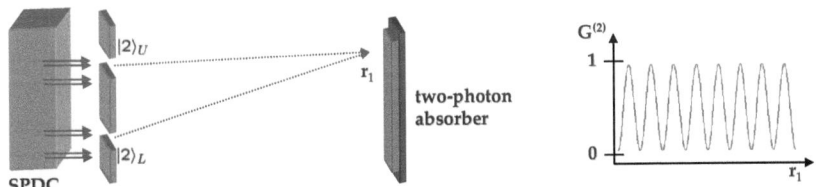

Figure 4.1: First experimental realization of N00N-state lithography for $N = 2$ in reference to [83]. The scheme uses entangled photon pairs generated by a non-linear crystal in the process of spontaneous parametric down conversion (SPDC). By arranging an double slit aperture directly behind the crystals surface, photons can pass by the double slit only in pairs where either both go through the upper or through the lower slit. The two photon absorber is realized by a beam splitter with two outputs and two avalanche photon diodes behind each output.

However, due to the Abbe criterion any sinusoidal pattern generated by a classical interferometry technique is limited to a fringe spacing of λ^1, where λ is the optical wavelength used. This limit could be overcome by using more complicated apertures, for example by the use of a grating. The smallest structures that can be obtained in an interference pattern of a grating aperture with N slits are of the size of λ/N. However, this pattern is no longer a single sinusoidal modulation and thus cannot be used to print *arbitrary* patterns.

In 2000 Boto *et al.* proposed how to *exploit entanglement to beat the diffraction limit* implementing quantum interferometry for purely lithographic purposes [33]. They consider the path-entangled N-photon state

$$|\psi(N)\rangle := 1/\sqrt{2}\,(|N\rangle_U|0\rangle_L + |0\rangle_U|N\rangle_L), \qquad (4.1.3)$$

which is also called a N00N-state, due to its structure [33].

An experimentally feasible setup capable of generating the photonic state $|\psi(N)\rangle$ for the case of $N = 2$ is illustrated in figure 4.1 which represents the first experimental realization by D'Angelo *et al.* [83] (see also [84]). Because the N-photon state $|\psi(N)\rangle$ has N-times the number of photons in a given mode than the single-photon state $|\psi(1)\rangle$, it accumulates an optical phase N-times faster as compared to $|\psi(1)\rangle$, when propagating through the setup. This gives rise to an N-photon absorption rate of the form

$$G^{(N)}(\mathbf{r},...,\mathbf{r}) \propto 1 + \cos N\delta(\mathbf{r}), \qquad (4.1.4)$$

---

[1]We note that this limit can be overcome by classical lithographic techniques. For example, using standing waves parallel to the substrate one can achieve a fringe spacing of $\lambda/2$. However, since we want to discuss the setup depicted in figure 4.1, its classical analog is limited by λ.

exhibiting a fringe spacing N-times narrower than that of $G^{(1)}(\mathbf{r})$ [33].

This gain in resolution can be fruitfully employed in a wide range of applications, e.g., lithography [33, 83, 84], microscopy [76], spectroscopy [88] and even magnetometry [89]. However, in order to implement this N-fold increase in resolution in an experiment, commonly an entangled state of the form $|\psi(N)\rangle$ in combination with a non-linear medium sensitive to N-photon absorption is needed [74]. These two prerequisites are usually in conflict with each other since multi-photon absorption requires high input fields whereas a high flux of N00N-states is increasingly difficult to realize for $N > 2$.

4.1.2 Example 2: N00N-state microscopy

As a further example, we briefly introduce the idea of using entangled photons for microscopic applications. In microscopy, we can make use of the Abbe criterion introduced in section 4.1 to determine the maximally obtainable resolution. When illuminating a small structure of interest some part of the light field will be scattered backwards. By detecting and analyzing this light it is possible to recover information about the scattering structure. However, if two light scattering sources are too close to each other, then the first principal maxima of the diffraction pattern of the backscattered light are no longer visible and hence the image of the sources will get blurred.

In 2004 Muthukrishnan et al. proposed how to *improve the microscopic resolution beyond the Rayleigh limit by using quantum light fields composed of entangled photons* [76] (see also [74, 90, 91]). In this method, as for N00N-state lithography, the authors consider the path-entangled number state $|\psi(N)\rangle$ (c.f. Eq. (4.1.3)). However, in difference to [33], the two-photon number state $|\psi(2)\rangle$ is generated by two atoms in a 3-level cascade emission as illustrated in figure 4.2.

In comparison with a classical setup like, e.g., the conventional Young double slit setup

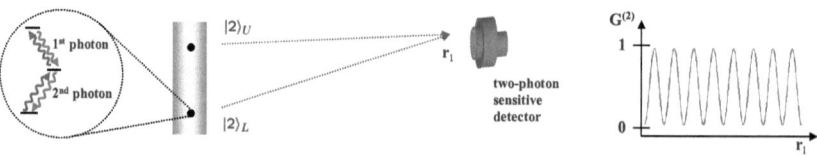

Figure 4.2: Illustration of N00N-state microscopy as proposed by [76]. The entangled photon number state $|\psi(2)\rangle$ is generated using atomic emitters: the detailed setup (c.f. [76]) ensures that only one of two possible atoms scatters two photons in a 3-level cascade emission. With a detector placed in the far-field region of the emitters and being sensitive to two-photon events only, there are only two possible modes $|2\rangle_U$ and $|2\rangle_L$ the photons can travel along.

which yields $G^{(1)}(\mathbf{r}) \propto 1 + \cos\delta(\mathbf{r})$, the N00N-state for $N=2$ gives rise to a two-photon absorption rate exhibiting a sinusoidal interference pattern with a fringe spacing two times narrower $G^{(2)}(\mathbf{r}) \propto 1 + \cos(2\,\delta(\mathbf{r}))$. In accordance with Abbe's criterion one thus has to scan only half of the range in the Fourier plane in order to detect the first diffraction order. In other words: for a fixed numerical aperture (assumed to be the same in both setups), this quantum microscopic method can resolve distances two times narrower than those which can be resolved by a classical microscope.

In order to implement this technique successfully, one has to guarantee that one and only one atom is excited, while keeping it impossible to determine which of the two atoms is in the excited state [76]. Subsequently, the excited atom will emit two photons in an atomic cascade. A two-photon absorbing detector is placed in the far-field region of the two atoms being able to record the spatial distribution of the two-photon state $|\psi(2)\rangle$ (c.f. figure 4.2). Therefore, the quantum imaging proposal of Muthukrishnan et $al.$ again implies the initial preparation of the two-photon state $|\psi(2)\rangle$ as well as a two-photon absorbing medium. The corresponding impediments have been discussed at the end of section 4.1.1. Nevertheless, in accordance with the Abbe criterion, the method can overcome the classical resolution limit by a factor of two.

4.2 A new ansatz for quantum imaging using incoherent photons

In the last section we reviewed and discussed two selected proposals in the context of quantum imaging. Representative for most approaches in this field, both proposals utilize path-entangled photon number states of the form $|\Psi(N)\rangle$ (c.f. Eq. (4.1.3)) and N-photon absorbers which allow to measure the N-photon absorption rate of the form $G^{(N)}(\mathbf{r},...,\mathbf{r}) \propto 1 + \cos N\delta(\mathbf{r})$ as discussed in Eq. (4.1.4). Contrary to these approaches and based on the work of Skornia et al. and subsequent investigations [60, 61, 77] we present a different ansatz for the implementation of quantum imaging in this section, using single trapped two-level atoms as a light source [19, 63].

We start to introduce the corresponding imaging scheme based on the setup of section 2.1. As it turns out this allows to achieve a resolution of λ/N, i.e., a resolution N-times higher than in a classical setup. Moreover, the scheme does not involve any of the requirements introduced in the foregoing section 4.1, i.e., it neither requires an initial N-photon number state $|\psi(N)\rangle$ nor relies on multi-photon absorbers. In what follows we will apply this scheme in the context of microscopy. The method employs N photons spontaneously emitted from N atoms and subsequently detected by all N detectors. This is ensured by means of post-selection (c.f. section 2.1.1). In this way, our approach employs only tools of linear optics as a single photon is registered at each detector. We demonstrate that in this case, for certain detector positions $\mathbf{r}_2,...,\mathbf{r}_N$, the intensity correlation signal of Nth order as a function of $\mathbf{r}_1$ takes the form $G^{(N)}(\mathbf{r}_1) \propto 1 + \cos N\delta(\mathbf{r}_1)$, resulting in a phase modulation with a theoretical contrast of 100% and a fringe spacing determined by λ/N. As with path-entangled number states, this corresponds to an N-fold reduced fringe spacing as compared to a classical intensity measurement of $G^{(1)}(\mathbf{r})$ in accordance with the Abbe criterion of section 4.1.

4.2.1 Model for quantum imaging based on Nth order correlations

To understand this outcome let us consider the basic setup of N identical two-level atoms excited by a single laser π-pulse as introduced in section 2.1. After their spontaneous emission, the N photons are registered by N detectors at positions $\mathbf{r}_1,...,\mathbf{r}_N$. For the sake of simplicity let us consider coincident detection. However, as we demonstrated in section 2.1.5 coincident detection is not a prerequisite for our scheme: the requirement is rather that all N photons emitted by the N atoms are recorded by the N distinct detectors; the exact detection time of a photon at a particular detector does neither influence the contrast nor the resolution of the correlation signal.

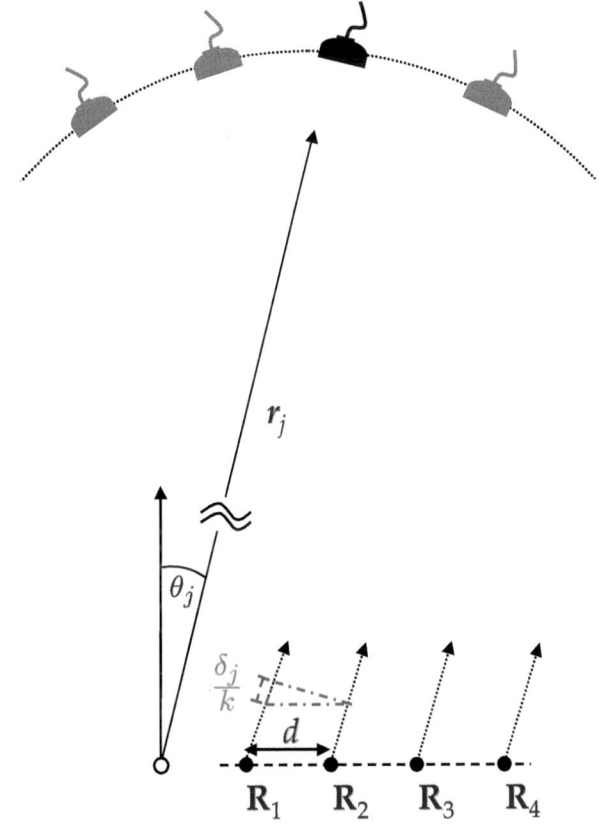

Figure 4.3: Arrangement of the atoms and detection scheme for the case of $N = 4$ identical two-level atoms at positions $\mathbf{R}_1, ..., \mathbf{R}_4$ which emit spontaneously four photons after excitation by a laser pulse. The photons are recorded in the far field by four detectors positioned at $\mathbf{r}_1, ..., \mathbf{r}_4$.

Figure 4.3 recalls the setup and the notation being used: N atoms at $\mathbf{R}_n$ ($n = 1, ..., N$) are aligned such that $\mathbf{R}_n = n\mathbf{R}_1$. In this case, we can make use of the optical phases $\delta_j = \delta(\mathbf{r}_j)$ as defined in Eq. (2.1.2), which depend solely on the location $\mathbf{r}_j$ of the jth detector. For the case of coincident detection the intensity correlation function of Nth order $G^{(N)}(\mathbf{r}_1, ..., \mathbf{r}_N)$ is defined in Eq. (2.2.15). The detection operator $\hat{D}_N(\delta_j)$ for the setup recalled in figure 4.3 is given by Eq. (2.2.14). As motivated in section 2.1, the arrangement of the system, i.e., the positions of the atoms $\mathbf{R}_n$ and the positions of the detectors $\mathbf{r}_j$, is important for the outcome of the correlation measurement.

Since all atoms are initially prepared in the excited state $\prod_{n=1}^{N} |e\rangle_n$ (c.f. Eq. (2.1.3)), we obtain from Eqs. (2.2.14) and (2.2.15)

$$G^{(N)}(\mathbf{r}_1, \ldots, \mathbf{r}_N) = \frac{\mathcal{E}_0^{2N}}{N^N} \left[\sum_{k=1}^{N!} \cos(\mathbf{j}_k \cdot \boldsymbol{\delta}) \right]^2. \quad (4.2.1)$$

Here, $\mathbf{j}_k$ is an N-component vector related to the equidistant spacing between the atoms

$$\mathbf{j}_k = P_k(1, \ldots, N), \quad (4.2.2)$$

where $\{P_k(1, ..., N)\}$ denotes the complete set of all $N!$ permutations of the tuple $(1, ..., N)$, while the vector $\boldsymbol{\delta}$ is given by:

$$\boldsymbol{\delta} = (\delta_1, \ldots, \delta_N). \quad (4.2.3)$$

The intensity correlation function of Nth order accounts for all possible $N!$ quantum paths, as the sum in Eq. (4.2.1) runs over all $N!$ permutations of the $\mathbf{j}_k$ components.

Due to the symmetry of the configuration (see figure 4.3), the intensity correlation function $G^{(N)}(\mathbf{r}_1, ..., \mathbf{r}_N)$ contains $N!/2$ elementary spatial frequencies in the arguments of the cosine functions of Eq. (4.2.1). Obviously, the complexity of the expression rises rapidly with increasing atom number N. However, if the N detectors are placed in such a manner that all terms in Eq. (4.2.1) interfere to give a single cosine, one is left with a modulation oscillating at a *unique* spatial frequency. This occurs in the following case [19, 63]:

- For arbitrary *even* N and choosing the detector positions such that

$$\begin{aligned}
\delta_2 &= -\delta_1, \\
\delta_3 &= \delta_5 = \ldots = \delta_{N-1} = \frac{2\pi}{N}, \\
\delta_4 &= \delta_6 = \ldots = \delta_N = -\frac{2\pi}{N},
\end{aligned} \quad (4.2.4)$$

the Nth order correlation function $G^{(N)}$ as a function of detector position $\mathbf{r}_1$ reduces to

$$G^{(N)}(\mathbf{r}_1) = A_N \left[1 + \cos(N\,\delta_1) \right], \quad (4.2.5)$$

where A_N is a dimensionless constant which depends on $\mathcal{E}_0$ and N.

- For arbitrary *odd* $N > 1$, and choosing the detector positions such that

$$\begin{aligned} \delta_2 &= -\delta_1, \\ \delta_3 &= \delta_5 = \ldots = \delta_N = \frac{2\pi}{N+1}, \\ \delta_4 &= \delta_6 = \ldots = \delta_{N-1} = -\frac{2\pi}{N+1}, \end{aligned} \quad (4.2.6)$$

the Nth order correlation function $G^{(N)}$ as a function of $\mathbf{r}_1$ reduces to

$$G^{(N)}(\mathbf{r}_1) = A_N \left[1 + \cos((N+1)\delta_1)\right]. \quad (4.2.7)$$

As demonstrated in Eqs. (4.2.5) and (4.2.7), we can construct a correlation signal of Nth order with a modulation of a single cosine, displaying a contrast of 100% and a fringe spacing determined by λ/N ($\lambda/(N+1)$) for any even (odd) N. We note that due to the limited detector sizes and the probabilistic ansatz implied by the use of spontaneously emitted photons only a subset of all emitted photons will be recorded. However, in contrast to those proposals based on maximally path-entangled N-photon states, in this scheme we are able to avoid both the necessity to generate a state of the form $|\psi(N)\rangle$ and the need to detect a multi-photon absorption signal[2] (c.f. section 4.1).

We emphasize that as the photons are produced by spontaneous decay the interference signal is generated by incoherent light. We stress further that the fringe contrast of 100% implied by Eq. (4.2.5) or Eq. (4.2.7) proves the underlying quantum nature of the process: we know that a classical light source can exhibit a two-photon signal with a contrast of 50% at best [59, 61, 92]. Moreover, it can be shown that if the contrast of the two-photon signal exceeds 71% it has the potential to violate Bell inequalities (c.f. chapter 6) This would prove ultimately the entangled nature of the correlations present in such a signal.

4.2.2 Examples for $N = 2$ and $N = 4$ emitters

To exemplify our method, let us consider the simplest situation, i.e., the case of $N = 2$ atoms, which is already known from section 3.1. With $\mathbf{j} = (1, 2)$ we obtain from Eq. (4.2.1)

$$G^{(2)}(\mathbf{r}_1, \mathbf{r}_2) = \frac{\mathcal{E}_0^4}{2} \left[1 + \cos(\delta_1 - \delta_2)\right], \quad (4.2.8)$$

which is plotted in figure 4.4 (a) - (c). Obviously, the modulation of the intensity correlation function of second order depends on the relative position of the two detectors: for the

---

[2]We want to point out that conditions (4.2.4) and (4.2.6) do not imply that some of the detectors have to be placed at the same position, since equality of the phase shifts may be obtained otherwise. For example, we could use the fact that the phase shifts are determined up to a multiple of 2π only.

choice of $\delta_2 = \delta_1$ the second order correlation function is a constant, whereas for fixed δ_2 the two photon counting rate as a function of δ_1 exhibits the same phase modulation and fringe spacing as the intensity signal $G^{(1)}(\mathbf{r})$ in Young's double slit experiment. However, the increased parameter space in terms of possible detector positions available in case of two detectors allows also to pick out the relative orientation $\delta_2 = -\delta_1$. In this case we get

$$G^{(2)}(\mathbf{r}_1) = \frac{\mathcal{E}_0^4}{2} \left[1 + \cos(2\delta_1)\right], \qquad (4.2.9)$$

exhibiting a phase modulation as a function of δ_1 with *half* the fringe spacing of $G^{(1)}(\mathbf{r})$ while keeping a contrast of 100% (see also [77]). Note that the assumed condition for the direction of emission of the two photons, i.e., $\delta_2 = -\delta_1$, corresponds to a space-momentum correlation of the photons identical to the one present in spontaneous parametric down conversion [83, 84, 86, 93].

In the case of the fourth order correlation function $G^{(4)}(\mathbf{r}_1, \mathbf{r}_2, \mathbf{r}_3, \mathbf{r}_4)$ for four equidistantly aligned atoms, placing the detectors according to Eq. (4.2.4), one finds

$$G^{(4)}(\mathbf{r}_1) = \frac{\mathcal{E}_0^8}{8} \left[1 + \cos(4\delta_1)\right], \qquad (4.2.10)$$

which is plotted in figure 4.4 (d). Obviously, $G^{(4)}(\mathbf{r}_1)$ as a function of δ_1 exhibits a modulation of a single cosine with a contrast of 100%. However, in this case the fringe spacing is determined by $\lambda/4$.

4.2.3 Quantum microscopy

As an example, let us apply our scheme in the context of microscopy. From Abbe's theory of the microscope introduced in section 4.1 we know that an object can be resolved only if at least the two first principal maxima of its diffraction pattern are visible in the Fourier plane [81].

Employing the first order correlation function $G^{(1)}(\mathbf{r}_1)$ for imaging N equidistant atoms equals the classical setup of imaging a regular grating with N slits using a coherent light source and measuring the intensity distribution in the far-field of the aperture. Indeed, if N atoms are initially prepared in a W-state (c.f. chapter 6)

$$|\phi\rangle = \frac{1}{\sqrt{N}} \sum_{i=1}^{N} \left(\prod_{n=1}^{i-1} |g\rangle_n \otimes |e\rangle_i \otimes \prod_{n=i+1}^{N} |g\rangle_n \right), \qquad (4.2.11)$$

where only one atom is excited, we get from Eqs. (2.2.14) and (2.2.15)

$$G^{(1)}(\mathbf{r}_1) = \frac{\mathcal{E}_0^2}{N} \left(1 + \sum_{n=1}^{N-1} (N-n) \cos(n\delta_1) \right), \qquad (4.2.12)$$

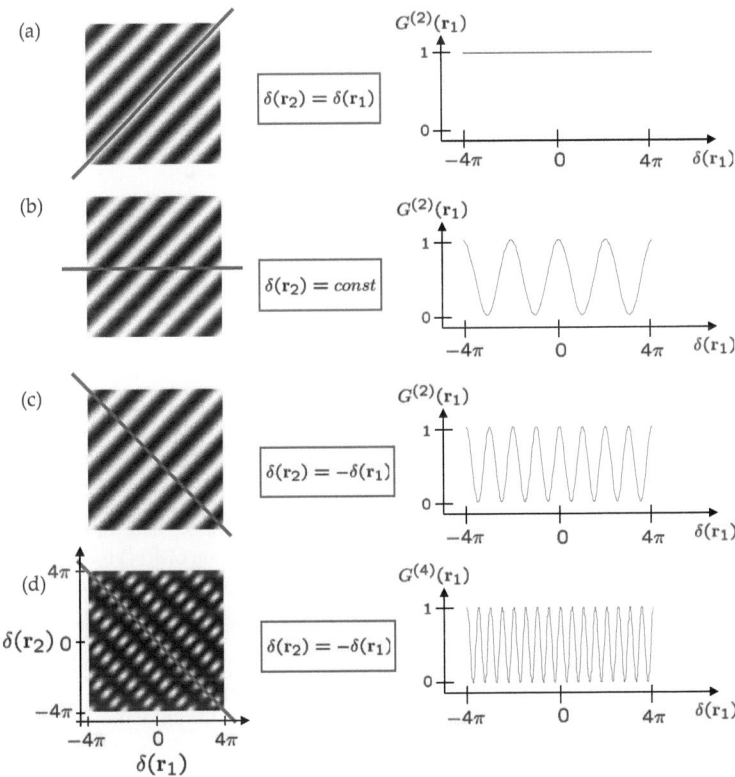

Figure 4.4: *Left* (a) - (c): density plots of $G^{(2)}(\mathbf{r}_1, \mathbf{r}_2)$ for two atoms versus $\delta(\mathbf{r}_1)$ and $\delta(\mathbf{r}_2)$; *left* (d): density plot of $G^{(4)}(\mathbf{r}_1, \mathbf{r}_2, \mathbf{r}_3, \mathbf{r}_4)$ for four atoms versus $\delta(\mathbf{r}_1)$ and $\delta(\mathbf{r}_2)$, with $\delta(\mathbf{r}_3) = \pi/2$ and $\delta(\mathbf{r}_4) = -\pi/2$.
Right: (normalized) cuts through the density plots along the indicated lines, i.e., for (a) $\delta(\mathbf{r}_2) = \delta(\mathbf{r}_1)$, (b) $\delta(\mathbf{r}_2) = const.$ and (c), (d) $\delta(\mathbf{r}_2) = -\delta(\mathbf{r}_1)$. All plots are normalized to unity.

which is the same intensity pattern as for the classical grating (c.f. section 4.3). Hereby, using Dirac's famous interpretation [29], it is sufficient to consider each photon independently to analyze the intensity signal, provided that a single photon can travel along any of the N different optical paths opened up by the N slits of the grating. Here, we prepare the N atoms in the W-state of Eq. (4.2.11) so that there are N possibly excited emitters where each may scatter a photon. However, since the detector in the far-field region cannot determine *which way* the photon traveled along, all quantum paths contribute equally to the measurement so that the $G^{(1)}(\mathbf{r}_1)$-function in Eq. (4.2.12) resembles the classical result obtained for the N-slit grating.

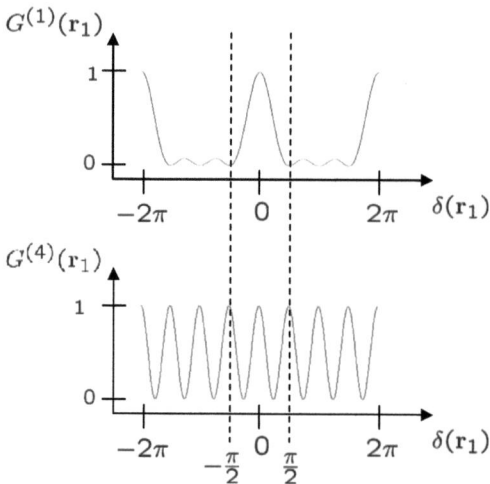

Figure 4.5: $G^{(1)}(\mathbf{r}_1)$ and $G^{(4)}(\mathbf{r}_1)$ as a function of $\delta(\mathbf{r}_1)$ for a chain of 4 atoms. The interval $[-2\pi, 2\pi]$ corresponds to the maximal range of variation of $\delta(\mathbf{r}_1)$ for an interatomic distance $d = \lambda$ (see Eq. (2.1.2)). The dashed lines indicate the corresponding range in case of $d = \lambda/4$. Both plots are normalized to unity.

As it is well-known for the grating equation (see Eq. (4.2.12)) and in accordance with Abbe's criterion two principal maxima appear in the far-field diffraction pattern only if it is possible to measure the intensity pattern in the range $-2\pi \leq kd\sin\theta \leq 2\pi$. In case of a numerical aperture[3] of $NA \cong 1$, this can be achieved only if the interatomic distance (generally the slit separation) is greater or equal to λ. In contrast, the use of the intensity correlation function of Nth order with the N detectors positioned according to Eq. (4.2.4) (or Eq. (4.2.6)) allows to resolve two principal maxima in the Fourier plane as long as it is possible to measure the intensity correlation function $G^{(N)}(\mathbf{r}_1)$ in the range $-2\pi \leq Nkd\sin\theta \leq 2\pi$. For $NA \cong 1$, this allows to resolve the atoms as long as their interatomic distance is greater or equal λ/N. For the case of $N = 4$, the two signals, namely the classical $G^{(1)}(\mathbf{r}_1)$-function and the intensity correlation function $G^{(4)}(\mathbf{r}_1)$, are compared in figure 4.5.

We want to point out that in the case of an interatomic spacing of $d \leq \lambda$ the intensity correlation function $G^{(N)}(\mathbf{r}_1)$ of Eq. (4.2.5) (or Eq. (4.2.7)) remains valid in case that one restricts the setup to coincident detection: in section 3.3 we demonstrated explicitly the influence of the dipole-dipole interaction on the spatial modulation of the intensity correlation function of second order. There we found that the spatial modulation of Eq. (4.2.5)

---

[3]The numerical aperture NA of a microscopic device describes its resolving power characterized by the range of observation angles over which the system can detect light.

can be retrieved even in the presence of the dipole-dipole interaction by restricting the measurement to coincident detection events (c.f. Eq. (3.3.5)).

4.2.4 Experimental feasibility and conclusions

Let us briefly address the technical feasibility of our quantum microscopy scheme. For the ability to localize atoms and adequately resolve optical path differences on a scale smaller than λ we refer to [94–96]. A detector of a given width s positioned at a distance $L = |\mathbf{r}_j|$ in the far-field region (see figure figure 4.3) gives rise to an angular resolution $\Delta\theta = s/L$, i.e., to a phase resolution $\Delta\delta = kd\cos\theta\Delta\theta$. To resolve the modulation of the N-th order correlation function $G^{(N)}(\mathbf{r}_1)$, a sufficient requirement is that $N\Delta\delta \ll 2\pi$, i.e., $\Delta\theta \ll \lambda/(Nd)$, which yields the condition

$$L \gg s\frac{Nd}{\lambda}. \qquad (4.2.13)$$

For given N, d and detector width s, we can thus always go to a distance L for which Eq. (4.2.13) is fulfilled so that the desired resolution is achieved. Hereby, choosing the smallest L compatible with Eq. (4.2.13) is favorable in order to maximize the N photon detection probability; the exact longitudinal positions of the detectors are thereby not important.

In case of a Gaussian distribution of the phases $\delta(\mathbf{r}_j)$ $(i = 2, \ldots, N)$ with standard deviation σ around their ideal values given by Eqs. (4.2.4) or (4.2.6) the contrast of the $G^{(N)}(\mathbf{r}_1)$-function is reduced to $e^{-N\sigma^2/4}$ [19, 25]. For $N = 2$ and $N = 4$, this means that a contrast higher than 50% can be maintained as long as σ is less than 0.8 and 1.2, respectively. Using the set of experimentally reasonable parameters $d = 5\,\mu\mathrm{m}$, $\Delta d = 0.1\,\mu\mathrm{m}$, $\theta(\mathbf{r}_1) = 30°$, $\Delta\theta(\mathbf{r}_1) = 0.1°$, $k = 2\pi/800\,\mathrm{nm}$, $\Delta k < 10^{-7}k$ we obtain $\sigma \approx 0.7$, which assures the required phase resolution for $N = 2$ and $N = 4$.

In section 4.2, we thus have shown that N photons of wavelength λ spontaneously emitted by N atoms and coincidently recorded by N detectors at particular positions exhibit correlations and interference properties similar to classical coherent light of wavelength λ/N. Thereby, the ansatz discussed provides a feasible scheme and allows to demonstrate an increase in resolution by the factor of N for a microscopic application due to quantum interference phenomena in comparison with a classical microscope. Moreover, the method requires neither initially entangled states nor multi-photon absorption, only commonly used single-photon detectors [19].

In analogy to the example of section 4.1.2, the image technique presented here is restricted insofar as the objects being imaged are the light emitting atoms themselves. However, so far all quantum imaging techniques known for the moment are incapable of imaging *arbitrary* objects with an enhanced spatial resolution (see for example section 4.1.1

and 4.1.2). By contrast, in the next section, we will investigate how our imaging ansatz can be extended to incorporate also the imaging of distinct physical objects, different from the source, such as, e.g., an aperture.

4.3 Quantum imaging of an aperture with sub-classical[4] resolution

In this section, we propose a method for imaging a *physical object*, e.g., an aperture, beyond the classical resolution limit using linear optics only. Again, we turn to the scheme introduced in section 2.1 which involves N uncorrelated single-photon emitters serving as a non-classical light source and N detectors performing the correlation measurements. By requiring again that all N detectors register exactly one photon within a measurement cycle and by placing the N detectors at different positions in the Fourier plane of the object we avoid the use of multi-photon absorption techniques. We exemplify our method for the case of two single-photon emitters. By exploiting two-photon interferences we show that this scheme allows to image a given object with sub-classical resolution, i.e., with a resolution enhanced by a factor of two with respect to the classical case. In the same way, sub-classical resolution enhanced by a factor of four is obtained for $N = 4$ emitters, using $N = 4$ detectors. By extending this scheme, we show that analogous results are also obtained for different objects, e.g., in case of a grating with N slits.

4.3.1 Description of the experimental configuration

The configuration of our quantum imaging scheme is shown in figure 4.6: two identical two-level atoms are located at $\mathbf{R}_1$ and $\mathbf{R}_2$ and initially excited by a single laser π pulse (where we denote the initially excited state by $|e, e\rangle := |e\rangle_1 \otimes |e\rangle_2$). After the spontaneous decay of both atoms, the two scattered photons are recorded by two detectors placed at $\mathbf{r}_1$ and $\mathbf{r}_2$ in the far-field region of the physical object which itself is placed between the atoms and the detection plane. In a first step we consider a rectangular aperture with opening height a and opening width b. In a single successful measurement cycle, the two photons emitted by the two atoms both pass through the aperture (distinct from *Ghost imaging* [70]; see also [97]) and are registered at the two detectors. For the sake of simplicity the two atoms and the two detectors each are assumed to be located in planes parallel to the x-y-plane, namely at $R_{1_z} = R_{2_z} =: R_z$ and $r_{1_z} = r_{2_z} =: r_z$.

Correlating the two detection events we measure the intensity correlation function of

---

[4]In the field of quantum imaging, and in particular when considering an image processing technique including an object, it became an Anglo-American tradition to speak of *sub-Rayleigh* resolution rather than referring to Abbe's definition. However, since we refer to Abbe's criterion of resolution (c.f. section 4.1), we make a compromise by speaking of a *sub-classical* resolution if a classical resolution which is determined by λ can be overcome, e.g., when a resolution of λ/N can be achieved (for arbitrary $N > 1$).

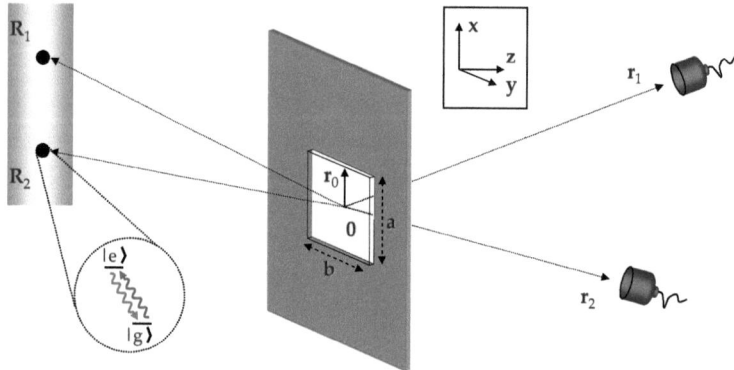

Figure 4.6: Setup used for far-field quantum imaging: two emitters at $\mathbf{R}_1$ and $\mathbf{R}_2$ each emit a single photon. Between the source and the detection plane, the light field is diffracted by an object (in the case depicted in this figure a rectangular aperture of height a and width b). The two detectors are placed in the far-field region of the aperture at positions $\mathbf{r}_1$ and $\mathbf{r}_2$. A measurement cycle is considered as successful if each of the two detectors registers a single photon.

second order (c.f. Eq. (2.2.5))

$$G^{(2)}(\mathbf{r}_1, \mathbf{r}_2) = \langle E^{(-)}(\mathbf{r}_1) E^{(-)}(\mathbf{r}_2) E^{(+)}(\mathbf{r}_2) E^{(+)}(\mathbf{r}_1) \rangle, \qquad (4.3.1)$$

which, again, can be understood as the joint probability to find one photon at $\mathbf{r}_1$ and another one at $\mathbf{r}_2$ (c.f. section 2.2.4). As we emphasized in section 2.2.3, we do not consider any explicit time dependence in the following analyses, as we may choose the atoms to be separated by an interatomic distance $d \gg \lambda$. In this case, we already demonstrated that the time dependence of intensity correlation functions of arbitrary order gives rise to an overall, exponentially decreasing term only which factorizes (c.f. Eq. (2.2.8)).

We note that Eq. (4.3.1) is formulated for the amplitudes of the electric field components only. As before we consider two-level atoms for the analyses of this section and thus we do not account explicitly for the polarization degrees of freedom.

4.3.2 Derivation of the disturbed electric field

From classical diffraction optics we know that the electric field amplitude $E(\mathbf{r})$ is diffracted at an aperture $\mathcal{A}$ placed between the source and the detection plane. The amplitude of the disturbed field $U(\mathbf{r})$ can be calculated employing *Fresnel-Kirchhoff* diffraction theory (see, e.g., [98]; a more detailed derivation can be found in appendix B). Using standard *Fresnel* approximations between source, aperture and detection plane we obtain the following expression for the diffracted field at $\mathbf{r}_j$ produced by a point source of unit strength at $\mathbf{R}_n$

$$U(\mathbf{r}_j, \mathbf{R}_n) = -\frac{iA}{\lambda} \frac{e^{ik|R_z|}e^{ik|r_z|}}{R_z r_z} \iint_{\mathcal{A}} e^{i\frac{k}{2}\frac{|\varrho_n-\rho_0|^2}{|R_z-r_{0_z}|}} e^{i\frac{k}{2}\frac{|\rho_0-\rho_j|^2}{|r_{0_z}-r_z|}} dS(\boldsymbol{\rho}_0). \tag{4.3.2}$$

Here, A denotes a constant proportional to the initial amplitude of the electric field at unit distance from the source, $k = \frac{2\pi}{\lambda}$ and $\mathbf{r}_0$ is a vector in the plane of the aperture where in particular $r_{0_z} = 0$ (see figure 4.6). $\boldsymbol{\varrho}_n$, $\boldsymbol{\rho}_0$ and $\boldsymbol{\rho}_j$ are vectors consisting of the x- and y-components of $\mathbf{R}_n$, $\mathbf{r}_0$ and $\mathbf{r}_j$ (with $n, j = 1, 2$), respectively. As one can see from Eq. (4.3.2), the problem is separated into the propagation of light from the source to the aperture (i.e., from $\mathbf{R}_n$ to $\mathbf{r}_0$) and further from the aperture to the detection plane (i.e., from $\mathbf{r}_0$ to $\mathbf{r}_j$). Since the far-field condition is assumed to be fulfilled in our setup we can restrict ourselves to the limit of *Fraunhofer*-diffraction, i.e.,

$$|R_z|, |r_z| \gg \frac{k}{2}|\boldsymbol{\rho}_0|^2, \tag{4.3.3}$$

(though the *Fresnel*-integral in Eq. (4.3.2) can be solved numerically without this approximation). In this limit, one can carry out the integral over $\mathcal{A}$ in Eq. (4.3.2) to obtain the final expression of the disturbed field

$$U(\mathbf{r}_j, \mathbf{R}_n) = \frac{iA\lambda R_z r_z}{\pi^2} e^{i\frac{k}{2}\frac{2R_z^2+|\varrho_n|^2}{R_z}} e^{i\frac{k}{2}\frac{2r_z^2+|\rho_j|^2}{r_z}} \tag{4.3.4}$$

$$\cdot \frac{\sin\left(\frac{kaR_{n_x}}{2R_z} + \frac{kar_{j_x}}{2r_z}\right)}{R_{n_x}r_z + r_{j_x}R_z} \cdot \frac{\sin\left(\frac{kbR_{n_y}}{2R_z} + \frac{kbr_{j_y}}{2r_z}\right)}{R_{n_y}r_z + r_{j_y}R_z}.$$

In figure 4.6, two atoms, i.e., two point-like sources, contribute to the electric field at $\mathbf{r}_j$, each giving rise to a disturbed field amplitude $U(\mathbf{r}_j, \mathbf{R}_n)$ of the form given in Eq. (4.3.4). We can thus write the total positive frequency part of the field contributing to the correlation signal at $\mathbf{r}_j$ as

$$E^{(+)}(\mathbf{r}_j) = \frac{1}{\sqrt{2}}\left(U(\mathbf{r}_j, \mathbf{R}_1)|g\rangle_1\langle e| + U(\mathbf{r}_j, \mathbf{R}_2)|g\rangle_2\langle e|\right), \tag{4.3.5}$$

where the atomic operator $|g\rangle_n\langle e|$ describes the de-excitation of the nth atom ($n = 1, 2$). Again, we make use of describing the detection of a single photon by projecting the respective emitter state back into its ground state as introduced in section 2.1.

Since the two atoms are initially in the state $|e, e\rangle$ and the electric field $E^{(+)}(\mathbf{r}_j)$ is given by Eq. (4.3.5), we can write the intensity correlation function of second order, Eq. (4.3.1), in the form

$$\begin{aligned} G^{(2)}(\mathbf{r}_1, \mathbf{r}_2) &= \left| E^{(+)}(\mathbf{r}_2) E^{(+)}(\mathbf{r}_1) |e, e\rangle \right|^2 \\ &= \frac{1}{4} \Big| \big(U(\mathbf{r}_j, \mathbf{R}_1) |g\rangle_1 \langle e| + U(\mathbf{r}_j, \mathbf{R}_2) |g\rangle_2 \langle e| \big) \\ &\quad \otimes \big(U(\mathbf{r}_j, \mathbf{R}_1) |g\rangle_1 \langle e| + U(\mathbf{r}_j, \mathbf{R}_2) |g\rangle_2 \langle e| \big) |e, e\rangle \Big|^2 \\ &= \frac{1}{4} \Big| U(\mathbf{r}_1, \mathbf{R}_1) U(\mathbf{r}_2, \mathbf{R}_2) + U(\mathbf{r}_1, \mathbf{R}_2) U(\mathbf{r}_2, \mathbf{R}_1) \Big|^2. \quad (4.3.6) \end{aligned}$$

Here, the last line of Eq. (4.3.6) displays the two quantum paths that contribute to the measurement: the photon emitted by the atom at $\mathbf{R}_1$ ($\mathbf{R}_2$) is either registered by the detector at $\mathbf{r}_1$ ($\mathbf{r}_2$) or by the one at $\mathbf{r}_2$ ($\mathbf{r}_1$).

4.3.3 Quantum imaging of a rectangular aperture with sub-classical resolution for $N = 2$ emitters

Utilizing Eqs. (4.3.4) and (4.3.6), it is possible to explicitly calculate the second order correlation function $G^{(2)}(\mathbf{r}_1, \mathbf{r}_2)$ for the setup shown in figure 4.6. For the sake of simplicity, let us consider the case where the detectors and the source, i.e., the emitters, are all coplanar in the x-z-plane, so that $r_{i_y} = R_{n_y} = 0$ ($n, j = 1, 2$). Moreover, we recall again that the determination of the intensity correlation function of second order $G^{(2)}(\mathbf{r}_1, \mathbf{r}_2)$, in contrast to the classical intensity pattern $I(\mathbf{r})$, requires the determination of *two* parameters when performing a measurement, namely $\mathbf{r}_1$ and $\mathbf{r}_2$. In addition, we have to fix the position of the two single-photon emitters with respect to the object which determines the phase shift between the different photon paths leading from either of the two emitters to the object (c.f. section 2.1).

In the case of a rectangular aperture of height a and width b and if we choose $R_{2_x} = R_{1_x} + \pi \frac{R_z}{ka}$ and $|r_{2_x}| = r_{1_x} := r_x$, we obtain for the intensity correlation function of second order along the x-axis

$$G^{(2)}(\mathbf{r}_1, \mathbf{r}_2) = \left(\frac{A^2 b^2 ka}{2\pi^2 R_z^2 B_\pm(r)} \right)^2 \cdot \sin^2 2\frac{kar_x}{2r_z}, \quad (4.3.7)$$

where $B_+(r_x) = r_x(\pi r_z + kar_x) \; [B_-(r_x) = r_x(\pi r_z - \frac{k^2 a^2 r_x^2}{\pi r_z})]$ holds for $r_{2_x} = +r_{1_x} \; [r_{2_x} = -r_{1_x}]$.

We note that the second order correlation signal of Eq. (4.3.6) consists of four contributions, each term oscillating generally with a different spatial frequency. However, using the set of parameters mentioned above all terms in Eq. (4.3.6) interfere to give a single sine (see Eq. (4.3.7)). In other words, one is left with a single sinusoidal modulation characterized by a *unique* spatial frequency.

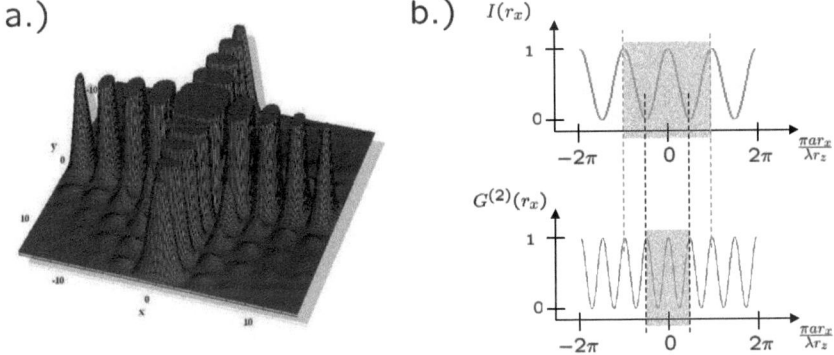

Figure 4.7: a.) Three dimensional plot of the classical diffraction pattern of a rectangular aperture. The illustration of Eq. (4.3.8) implies arbitrary parameters (the intensity is cut for a clearer picture). b.) Comparison of the classical intensity pattern $I(r_x)$ of Eq. (4.3.8) and the second order function $G^{(2)}(r_x)$ of Eq. (4.3.7); both plots imply arbitrary parameters are normalized to unity.

To clarify this point, let us briefly describe the classical analog of our setup and compare the resolution obtained in both cases: in classical optics, using a coherent light source and in the limit of Fraunhofer-diffraction, it is well-known that a rectangular aperture with opening height a and width b gives rise to the following classical intensity diffraction pattern at a point $\mathbf{r}$ in the far field (see [98])

$$I(\mathbf{r}) = \left(\frac{8Ar_z}{\pi k r_x r_y R_z}\right)^2 \cdot \sin^2 \frac{kar_x}{2r_z} \cdot \sin^2 \frac{kbr_y}{2r_z}, \qquad (4.3.8)$$

where R_z denotes the distance between the source and the aperture. Comparing the modulation of the classical intensity pattern $I(\mathbf{r})$ from Eq. (4.3.8) along the x-axis with the one obtained for the $G^{(2)}$-function in Eq. (4.3.7), we find that the latter oscillates twice as fast as in the classical case due to a useful exploitation of the increased parameter space available (c.f. also figure 4.7). According to the Abbe criterion introduced in section 4.1, this increase of the modulation frequency by a factor of two implies that sufficient information is available in the Fourier plane to reconstruct the aperture even if measuring only half of the Fourier range needed for the classical imaging technique. For a given NA, the object can thus be imaged with a resolution enhanced by a factor of two.

We note that our scheme also allows to reproduce the object in the image plane of a lens placed in the Fourier plane of the object. So far, we assumed that the two detectors scan the range in the Fourier plane at *different* positions, $r_{2_x} = -r_{1_x}$, so that the joint detection measurements can be performed using ordinary single-photon detectors (see figure 4.8a).

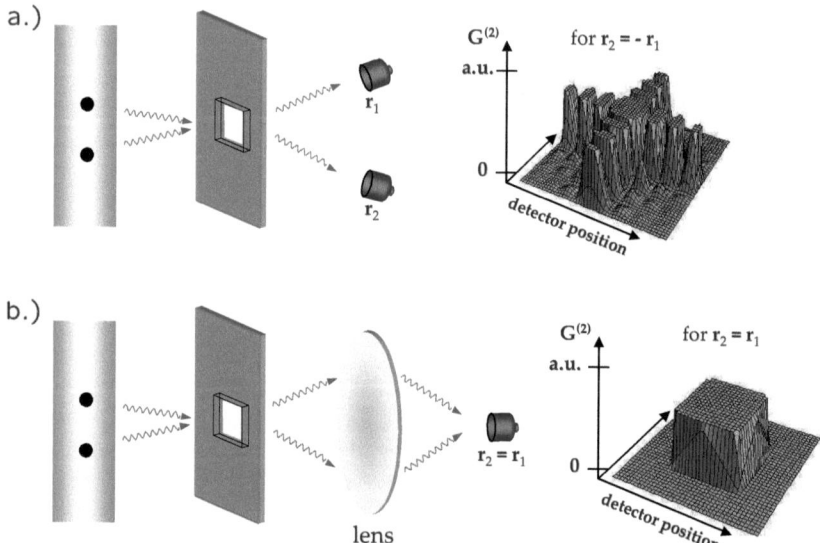

Figure 4.8: *a.*) Enhanced two-photon imaging setup employing different detector positions. *b.*) By using a lens in the Fourier plane of the object and a two-photon absorbing medium our scheme is able to reconstruct the object with sub-classical resolution without relying on a post selection mechanism.

However, in order to create an image of the object in the image plane of the lens, we have to relocate the two detectors at the *same* position in the image plane as shown in figure 4.8b. Hereby, it is convenient that $|r_{2_x}| = r_{1_x}$ is compatible with $r_{2_x} = -r_{1_x}$ and $r_{2_x} = r_{1_x}$ (see Eq. (4.3.7)): the different quantum paths associated with the two scenarios cannot be distinguished in the image plane and thus they all contribute to each measurement (see figure 4.8b). By using, e.g., a single detector sensitive to two-photon absorption it becomes possible to reconstruct the object with sub-classical resolution in the image plane. We note that by using a two-photon absorbing medium our scheme does not rely on a post selection mechanism anymore; in this way the scheme becomes also relevant for lithographic applications, too (see, e.g., [63]).

4.3.4 Quantum imaging of a rectangular aperture with sub-classical resolution for $N = 4$ emitters

The results found in section 4.3.3 for the case of $N = 2$ emitters using $N = 2$ detectors can be extended to the case of $N > 2$. For example, if we consider $N = 4$ single-photon emitters located at positions $R_{1_x} = -\pi \frac{R_z}{ka}$, $R_{2_x} = 0$, $R_{3_x} = \frac{\pi}{2} \frac{R_z}{ka}$, $R_{4_x} = \pi \frac{R_z}{ka}$ and choose the positions $|r_{2_x}| = r_x$, $r_{3_x} = -r_x + \pi \frac{r_z}{ka}$, $r_{4_x} = r_x + \frac{\pi}{2}\frac{r_z}{ka}$ for the four detectors, where again $r_{1_x} =: r_x$, we obtain the following expression for the intensity correlation function of fourth order along the x-axis (see section 2.2.3)

$$G^{(4)}(\mathbf{r}_1,\mathbf{r}_2,\mathbf{r}_3,\mathbf{r}_4) \propto \sin^2 4\frac{kar_x}{2r_z}. \qquad (4.3.9)$$

As before, we find that it is possible to image the object with sub-classical resolution. However, using four emitters and four detectors, the resolution is now enhanced by a factor of four with respect to the classical case (compare with Eq. (4.3.8)).

4.3.5 Quantum imaging of a grating with M slits and sub-classical resolution for $N = 2$ emitters

Finally, we briefly outline that our method can be extended to image different objects, e.g., to the case of an arbitrary grating. Therefore, let us reconsider the expression of the disturbed field $U(\mathbf{r}_j, \mathbf{R}_{n_z})$ of a single rectangular aperture as derived in Eq. (4.3.4). In the case of two emitters and a grating with M slits (opening height a, width b and slit separation d), each photon may pass through either of the M slits before being recorded by one of the two detectors at $\mathbf{r}_1$ or $\mathbf{r}_2$.

For simplicity, again, we assume emitters, aperture and detectors to be located coplanar and restrict our calculations to the x-z-plane. In this case, the general expression of the electric field being diffracted at a grating with M slits is given by

$$\begin{aligned} U(\mathbf{r}_j, \mathbf{R}_n, M) &= U(\mathbf{r}_j, \mathbf{R}_n) \cdot \sum_{m=0}^{M-1} e^{ikmd\frac{R_{n_x}}{R_{n_z}}} \cdot \sum_{m=0}^{M-1} e^{-ikmd\frac{r_{j_x}}{r_z}} \\ &= U(\mathbf{r}_j, \mathbf{R}_n) \cdot \frac{1 - e^{ikMd\frac{R_{n_x}}{R_{n_z}}}}{1 - e^{ikd\frac{R_{n_x}}{R_{n_z}}}} \cdot \frac{1 - e^{-ikMd\frac{r_{j_x}}{r_z}}}{1 - e^{-ikd\frac{r_{j_x}}{r_z}}}, \end{aligned} \qquad (4.3.10)$$

where, again, we made use of Fresnel and Fraunhofer approximations. Using this expression and choosing $|r_{2_x}| = r_{1_x} \pm \frac{\pi r_z}{kd}$, the second order correlation function $G^{(2)}(\mathbf{r}_1, \mathbf{r}_2, M)$ for a grating with an odd number M of slits is given by

$$G^{(2)}(\mathbf{r}_1, \mathbf{r}_2, M) = G^{(2)}(\mathbf{r}_1, \mathbf{r}_2, 1) \cdot \frac{1 - \cos[k2Md\frac{r_{1_x}}{r_z}]}{1 - \cos[k2d\frac{r_{1_x}}{r_z}]}, \qquad (4.3.11)$$

where $G^{(2)}(\mathbf{r}_1, \mathbf{r}_2, 1) := G^{(2)}(\mathbf{r}_1, \mathbf{r}_2)$ is the second order correlation signal for a single slit aperture (c.f. Eq.(4.3.7)).

The classical expression for the intensity diffraction pattern of a grating with M slits in case of a coherent source is well known [98]

$$I(\mathbf{r}, M) = I(\mathbf{r}, 1) \cdot \frac{1 - \cos[kMd\frac{r_x}{r_z}]}{1 - \cos[kd\frac{r_x}{r_z}]}, \qquad (4.3.12)$$

where $I(\mathbf{r}, 1) = I(\mathbf{r})$ is the intensity diffraction pattern of a single slit (c.f. Eq.(4.3.8)). Comparing Eq. (4.3.11) with (4.3.12), we see that both expressions can be written as a product of an envelope function due to the diffraction at a single slit ($G^{(2)}(\mathbf{r}_1, \mathbf{r}_2, 1)$ and $I(\mathbf{r}, 1)$, respectively) and a sinusoidally oscillating function. However, in case of the $G^{(2)}$-function the modulation frequency is again twice as high as in the analogous classical case indicating an imaging with a sub-classical resolution, i.e., with a resolution enhanced by a factor of two.

4.4 Conclusions: a comparison with experiment

Summarizing, the setup proposed in section 4.3 can be used to image a physical object with sub-classical resolution using uncorrelated single-photon emitters as a source and tools of linear optics only [23]. In section 4.2, it was demonstrated how the *source* of single-photon emitters itself can be imaged and resolved with sub-classical resolution using ordinary photon detectors and joint detection techniques (see also [19]). In contrast, the scheme developed in section 4.3 is able to resolve details of a *distinct physical object*, e.g., a rectangular aperture or a periodical structure, with a resolution impossible to achieve with classical far-field imaging techniques [23]. We remark that our method can be implemented in various physical systems, e.g., with current ion trap technology [8,9,41,42].

Let us conclude by comparing our ansatz for quantum imaging discussed throughout chapter 4 with the one by D'Angelo *et al.* which was demonstrated in an experiment in 2001 [73]. In this experiment the authors achieved a higher spatial resolution when imaging a double slit as attainable with classical imaging. They used pairs of entangled photons generated by the process of parametric down-conversion as suggested in [33] (see also figure 4.1). Here, interferences of two-photon amplitudes of the kind $|2\rangle_U \otimes |0\rangle_L + |0\rangle_U \otimes |2\rangle_L$ are exploited while both photons are registered at the same position (see fig. 4.9a). Later, it was shown by the same research group that entangled photons are not necessary to overcome the classical resolution limit, since the same interference pattern - though with reduced contrast - was observed when using thermal light at low intensity [99]. In the respective experiment, two detectors register the two-photon events at different positions ($r_{2_x} = -r_{1_x}$) so that only the state $|1\rangle_U \otimes |1\rangle_L$ of the initial Fock-state $|2\rangle_U \otimes |0\rangle_L + |0\rangle_U \otimes |2\rangle_L + |1\rangle_U \otimes |1\rangle_L$ contributes to the interference pattern with doubled spatial modulation frequency while the states $|2\rangle_U \otimes |0\rangle_L$ and $|0\rangle_U \otimes |2\rangle_L$ give rise to an unwanted background signal reducing the contrast to 33% only (see fig. 4.9b).

In contrast to the experiment of D'Angelo *et al.* [73], our scheme uses initially uncorrelated, incoherent photons (see fig. 4.9c). Nevertheless, it enables us to image a physical object with a contrast of 100%. In analogy with [99] our initial photon state, contributing to the diffraction image, can be written as $|1\rangle_U \otimes |1\rangle_L$ where the two modes U and L are determined by the positions of the two emitting atoms at $\mathbf{R}_1$ and $\mathbf{R}_2$, respectively. The reason why the contrast of our correlation signal is not reduced due to any disturbing contributions is naturally attributed to the non-classicality of our source: using two single-photon emitting atoms as source our scheme enables us to select solely the pure Fock-state $|1\rangle_U \otimes |1\rangle_L$.

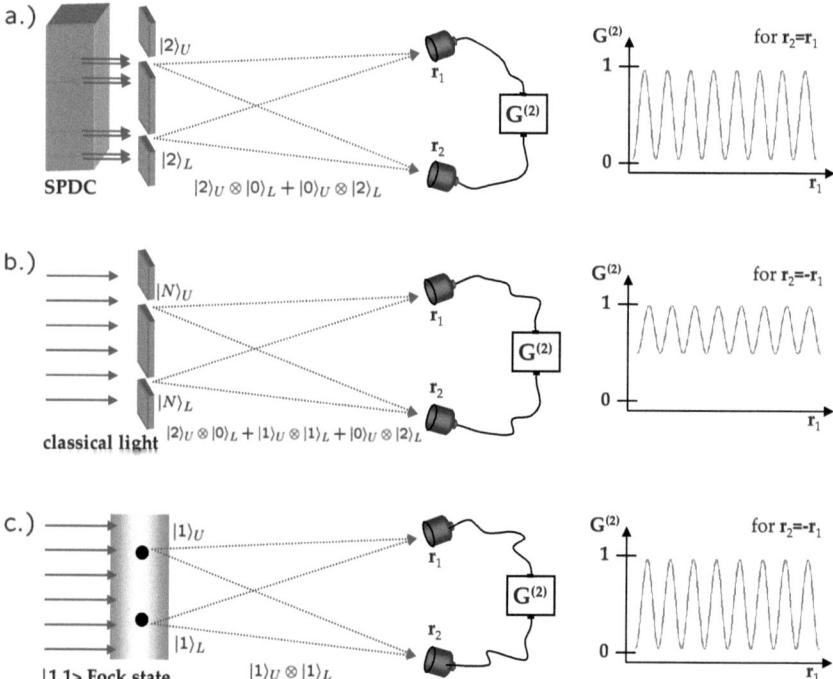

Figure 4.9: *a.*) Two-photon quantum imaging using initially entangled photons via spontaneous parametric down conversion (SPDC) as proposed by [33]. *b.*) It was shown [99] that two-photon quantum imaging can enhance the resolution by using incoherent photons, too, but with a contrast of 33% at best. *c.*) By using a non-classical source of uncorrelated photons one can enhance the resolution but remain with a contrast of 100% (here two atoms generate initially a photonic Fock-state $|1\rangle_U \otimes |1\rangle_L$).

Chapter 5

Quantum state engineering

In the first part of this thesis the focus of research was mainly restricted to the analysis and application of the fluorescence signal of the atomic emitters. Different orders of correlations were investigated and utilized for the purpose of quantum imaging, i.e., an image processing technique able to overcome classical resolution limits such as the Rayleigh- or the Abbe-limit [80, 81]. Quantum imaging exploits fundamental quantum characteristics such as higher order correlations by means of experimentally easily accessible parameters namely the detector positions. In this context it is sufficient to use two-level atoms, being excited and decaying within a closed measurement cycle, as a source for indistinguishable photons.

The present chapter will extend these considerations. In particular, we will investigate the back action of measuring the fluorescence photons on the quantum state of the emitting atoms. For this purpose we will use a more complex atomic level-structure, namely a Λ-level structure, characterized by one excited and two ground states. Using N atoms with a Λ-level structure this setup provides a natural N-qubit basis among the ground states. By modifying the geometry of the system, resembling the way correlation signals were manipulated in chapter 4, and by accessing further degrees of freedom as, e.g., the photons' polarization state, we investigate how the final projected state of the N atoms can be manipulated.

In the first part of this chapter we introduce the modified system, detection operators and qubit basis. Thereafter, we consider slightly different implementations of our basic setup that enable the generation of different classes of multi-partite entangled quantum states within the long-lived N-qubit basis [21, 24]. We will show that the scheme used here allows to classify these quantum states by experimentally easily accessible parameters [22].

5.1 Introduction to quantum state engineering

5.1.1 Atom-photon entanglement

In 2004 Blinov et al. observed entanglement between a single trapped ^{111}Cd$^+$ ion and a spontaneously emitted single photon [100]. The ^{111}Cd$^+$ ion is characterized by a Λ-level structure, i.e., a level structure where the excited state $|e\rangle$ can decay along two different transitions

$$|e\rangle \xrightarrow{\epsilon_1} |g_1\rangle \quad \text{and} \quad |e\rangle \xrightarrow{\epsilon_2} |g_2\rangle, \tag{5.1.1}$$

to either of the two ground states $|g_1\rangle$ or $|g_2\rangle$ by emitting a photon with polarization ϵ_1 or ϵ_2, respectively, while obeying the conservation of angular momentum of the composite atom-photon system. The state of the composite atom-photon system $|\Phi\rangle$, after emission of the photon but before any measurement is performed, is a superposition of two possible quantum states corresponding to the two quantum paths the system can evolve along. In case both transitions are equally probable, both quantum states are occupied with equal probabilities and thus the state of the composite system $|\Phi\rangle$ is given by

$$|\Phi\rangle = \frac{1}{\sqrt{2}}(|g_1\rangle|\epsilon_1\rangle + |g_2\rangle|\epsilon_2\rangle). \tag{5.1.2}$$

We note that the state $|\Phi\rangle$ is a maximally entangled 2-qubit state, as can be seen, e.g., by calculating the entanglement of formation as described in [101].

The atom-photon entanglement of the state $|\Phi\rangle$ provides us with one of two important *tools* essential for the scheme of measurement based quantum state engineering that will be introduced in the following. It enables us to manipulate the ground state population of the emitters by utilizing the photonic polarization degree of freedom. The second *tool* is, again, provided by a far-field or fiber-based detection scheme which does not allow to obtain any *which-way* information due to the data obtained in a measurement (c.f. section 2.1).

5.1.2 Description of the physical system employing emitters with Λ-level structure

We consider a system of N single-photon emitters with a Λ-level structure as depicted in figure 5.1. The two possible transitions are described in Eq. (5.1.1). To simplify all further investigations, we restrict ourselves to a system where both transitions are equally probable. As before, atoms localized in an ion trap provide a setup can be controlled to a very high degree [8, 9, 41, 42].

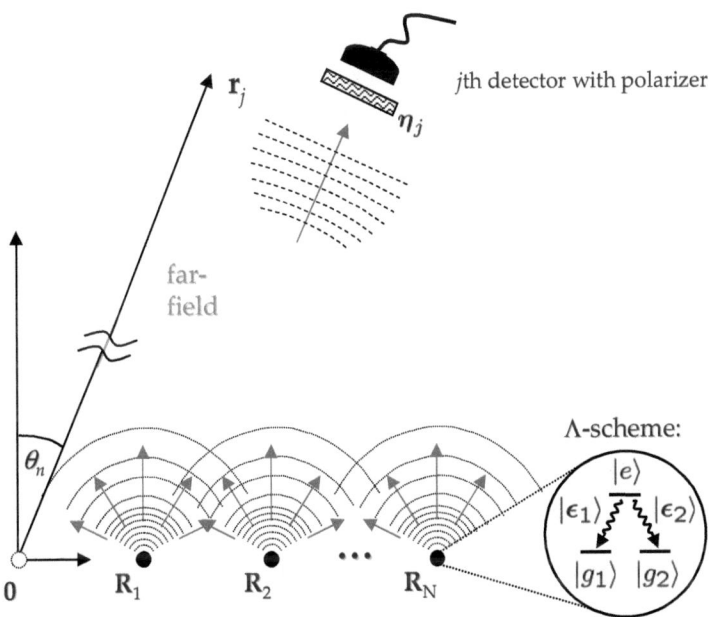

Figure 5.1: N atomic emitters are regularly aligned in a row with spacing d. The definition of the coordinate system is taken over from figure 2.3. The jth detector is placed at $\mathbf{r}_j$ in the far-field region of the atoms, where all atoms appear under an angle θ_j with respect to a line perpendicular to the symmetry axis of the alignment. The jth detector is further equipped with a polarization filter transmitting only light of polarization $\boldsymbol{\eta}_j$.

We define a similar measurement cycle as in chapter 4: initially all N emitters are excited into the upper state $|e\rangle$ by a laser π pulse. Here, we denote the initial state of all N atoms of Eq. (2.1.3) by $|\Psi_N^i\rangle$ which is thus given by

$$|\Psi_N^i\rangle := \prod_{n=1}^{N} |e\rangle_n = |e, e, \ldots, e\rangle_N, \qquad (5.1.3)$$

where the dimension of the state is indicated by the subscript N. Subsequently, the N emitters scatter N photons which are to be registered at N distinct detectors. Via post-selection, only those events where all detectors register one and only one photon will be accepted as a successful measurement. Eventually, all N atoms have thus been projected into a ground state. In contrast to all former investigations, the detection is performed polarization sensitive, enabling us to *project* certain emitter states, as will be demonstrated in the following.

In accordance with the setup introduced in the previous chapters, the N atomic emitters with a Λ-level structure are arranged in a regular line (c.f. figure 5.1). Again, the coordinate system is chosen such that $\mathbf{R}_n = n\,\mathbf{R}_1$ for $n = 1, ..., N$. The N detectors are located in the far-field region of the emitters at positions $\mathbf{r}_j$. In difference to the former setup, they are now equipped with polarization sensitive filters that transmit only polarized light with a polarization vector pointing in $\boldsymbol{\eta}_j$ direction ($j = 1, ..., N$). Extending Eq. (2.2.13) by the novel specifications of our setup, the detection operator describing a photon event at $\mathbf{r}_j$ is given by

$$\hat{D}_N(\delta_j, \boldsymbol{\eta}_j) = \frac{\mathcal{E}_j}{\sqrt{N}} \sum_{n=1}^{N} \left(e^{in\delta_j} \left(\boldsymbol{\eta}_j \cdot \boldsymbol{\epsilon}_1 \right) |g_1\rangle_n \langle e| + e^{in\delta_j} \left(\boldsymbol{\eta}_j \cdot \boldsymbol{\epsilon}_2 \right) |g_2\rangle_n \langle e| \right), \quad (5.1.4)$$

where δ_j is defined as in Eq. (2.1.2) and $\mathcal{E}_j$ is a redefined field amplitude given by

$$\mathcal{E}_j^2 := \mathcal{E}_0^2 / (|(\boldsymbol{\eta}_j \cdot \boldsymbol{\epsilon}_1)|^2 + |(\boldsymbol{\eta}_j \cdot \boldsymbol{\epsilon}_2)|^2). \quad (5.1.5)$$

We note that this operator accounts explicitly for the polarization degree of freedom.

5.1.3 A measurement scheme based on *projection*

In order to understand the concept of *projection* of quantum states by a measurement process let us briefly recall the most basic process present in our system: if one single excited atom with a Λ-level structure emits a photon, the atomic ground state and the photonic polarization state cannot be described independently. The excited state $|e\rangle$ can decay along the two possible channels described in Eq. (5.1.1), accompanied by the spontaneous emission of a photon polarized along $\boldsymbol{\epsilon}_1$ or $\boldsymbol{\epsilon}_2$, respectively (consider, e.g., Zeeman sub-levels). The atomic ground state and the corresponding photonic polarization state associated with a single decaying atom and its scattered photon thus form the entangled state $|\Phi\rangle$ of Eq. (5.1.2) [44, 100].

This correlation implies that the state of an atom with a Λ-level structure can be projected onto $|g_1\rangle$ ($|g_2\rangle$) if the emitted photon is registered by a detector with a polarization filter in front transmitting only $\boldsymbol{\epsilon}_1$ ($\boldsymbol{\epsilon}_2$) polarized light. In other words, the polarization sensitive detection of a photon entangled with an emitting atom as described by Eq. (5.1.2) can be used to manipulate or to *project* a particular ground state of the atomic emitter. The latter provides a *long-lived* qubit basis (formed by the two different ground states) in terms of time scales relevant to the quantum optics community that uses entangled states for various applications in quantum information science.

Let us exemplify our method for $N = 2$ atomic emitters with a Λ-level structure. As before, the detection of the two photons can be described by means of the intensity correlation function of second order. While the detection of a photon is connected with the projection of the corresponding emitter state, the correlation function is also able to describe the successive evolution of the atomic state itself.

The two atomic emitters with a Λ-level structure are located at $\mathbf{R}_1$ and $\mathbf{R}_2$ and two detectors equipped with polarization sensitive filters that transmit only $\boldsymbol{\eta}_1$ and $\boldsymbol{\eta}_2$ polarized light are placed in the far-field region of the emitters at $\mathbf{r}_1$ and $\mathbf{r}_2$, respectively. Using Eqs. (2.2.15) and (5.1.4), we can determine the intensity correlation function of second order $G^{(2)}(\mathbf{r}_1, \mathbf{r}_2, \boldsymbol{\eta}_1, \boldsymbol{\eta}_2)$ given by

$$\begin{aligned} G^{(2)}(\mathbf{r}_1, \mathbf{r}_2, \boldsymbol{\eta}_1, \boldsymbol{\eta}_2) &= \left| \hat{D}_2(\mathbf{r}_2, \boldsymbol{\eta}_2)\, \hat{D}_2(\mathbf{r}_1, \boldsymbol{\eta}_1) |e, e\rangle \right|^2 \\ &= \frac{\mathcal{E}_1^2}{2} \big| \hat{D}_2(\mathbf{r}_2, \boldsymbol{\eta}_2) \big(e^{i\delta_1} (\boldsymbol{\eta}_1 \cdot \boldsymbol{\epsilon}_1) |g_1, e\rangle + e^{i\delta_1} (\boldsymbol{\eta}_1 \cdot \boldsymbol{\epsilon}_2) |g_2, e\rangle \\ &\quad + e^{i2\delta_1} (\boldsymbol{\eta}_1 \cdot \boldsymbol{\epsilon}_1) |e, g_1\rangle + e^{i2\delta_1} (\boldsymbol{\eta}_1 \cdot \boldsymbol{\epsilon}_2) |e, g_2\rangle \big) \big|^2 . \end{aligned} \qquad (5.1.6)$$

We note that our notation of the correlation function used here and in the following is extended by the polarization degrees of freedom in accordance with the detection operator of Eq. (5.1.4). So far, in Eq. (5.1.6) we applied explicitly only the first detection operator. Nevertheless, we see immediately that by substituting the two-level emitters employed in the previous chapters with emitters which exhibit a Λ-level structure, the number of quantum paths the system can evolve along is increased significantly.

Proceeding with the calculation of the second order correlation function we find

$$\begin{aligned} G^{(2)}(\mathbf{r}_1, \mathbf{r}_2, \boldsymbol{\eta}_1, \boldsymbol{\eta}_2) =& \\ \frac{\mathcal{E}_1^2 \mathcal{E}_2^2}{4} \big| & e^{i\delta_1 + i 2\delta_2} (\boldsymbol{\eta}_1 \cdot \boldsymbol{\epsilon}_1)(\boldsymbol{\eta}_2 \cdot \boldsymbol{\epsilon}_1) |g_1, g_1\rangle + e^{i\delta_1 + i 2\delta_2} (\boldsymbol{\eta}_1 \cdot \boldsymbol{\epsilon}_1)(\boldsymbol{\eta}_2 \cdot \boldsymbol{\epsilon}_2) |g_1, g_2\rangle \\ +& \; e^{i\delta_1 + i 2\delta_2} (\boldsymbol{\eta}_1 \cdot \boldsymbol{\epsilon}_2)(\boldsymbol{\eta}_2 \cdot \boldsymbol{\epsilon}_1) |g_2, g_1\rangle + e^{i\delta_1 + i 2\delta_2} (\boldsymbol{\eta}_1 \cdot \boldsymbol{\epsilon}_2)(\boldsymbol{\eta}_2 \cdot \boldsymbol{\epsilon}_2) |g_2, g_2\rangle \\ +& \; e^{i 2\delta_1 + i\delta_2} (\boldsymbol{\eta}_1 \cdot \boldsymbol{\epsilon}_1)(\boldsymbol{\eta}_2 \cdot \boldsymbol{\epsilon}_1) |g_1, g_1\rangle + e^{i 2\delta_1 + i\delta_2} (\boldsymbol{\eta}_1 \cdot \boldsymbol{\epsilon}_1)(\boldsymbol{\eta}_2 \cdot \boldsymbol{\epsilon}_2) |g_2, g_1\rangle \\ +& \; e^{i 2\delta_1 + i\delta_2} (\boldsymbol{\eta}_1 \cdot \boldsymbol{\epsilon}_2)(\boldsymbol{\eta}_2 \cdot \boldsymbol{\epsilon}_1) |g_1, g_2\rangle + e^{i\delta_1 + i\delta_2} (\boldsymbol{\eta}_1 \cdot \boldsymbol{\epsilon}_2)(\boldsymbol{\eta}_2 \cdot \boldsymbol{\epsilon}_2) |g_2, g_2\rangle \big|^2 . \end{aligned} \qquad (5.1.7)$$

By inspection of Eq. (5.1.7) we may learn how the intensity correlation measurement determines the final state of the atomic emitters: depending on terms of the form $\boldsymbol{\eta}_j \cdot \boldsymbol{\epsilon}_n$ ($j, n = 1, 2$), the final state of the atomic emitters is given by a superposition of all four two-qubit basis states $|g_1, g_1\rangle$, $|g_1, g_2\rangle$, $|g_2, g_1\rangle$ and $|g_2, g_2\rangle$.

Before we consider the final state of the atomic emitters in more detail, we continue with our calculation of $G^{(2)}(\mathbf{r}_1, \mathbf{r}_2, \boldsymbol{\eta}_1, \boldsymbol{\eta}_2)$. Simplifying Eq. (5.1.7) we obtain

$$\begin{aligned} G^{(2)}(\mathbf{r}_1, \mathbf{r}_2, \boldsymbol{\eta}_1, \boldsymbol{\eta}_2) =& \\ \frac{\mathcal{E}_1^2 \mathcal{E}_2^2}{4} \big[& (2 + 2\cos(\delta_2 - \delta_1)) \left(|(\boldsymbol{\eta}_1 \cdot \boldsymbol{\epsilon}_1)(\boldsymbol{\eta}_2 \cdot \boldsymbol{\epsilon}_1)|^2 + |(\boldsymbol{\eta}_1 \cdot \boldsymbol{\epsilon}_2)(\boldsymbol{\eta}_2 \cdot \boldsymbol{\epsilon}_2)|^2 \right) \\ +& (2 + 2\cos(\delta_2 - \delta_1)) \left| (\boldsymbol{\eta}_1 \cdot \boldsymbol{\epsilon}_1)^\dagger (\boldsymbol{\eta}_2 \cdot \boldsymbol{\epsilon}_2)^\dagger (\boldsymbol{\eta}_1 \cdot \boldsymbol{\epsilon}_2)(\boldsymbol{\eta}_2 \cdot \boldsymbol{\epsilon}_1) \right| \\ +& (2 + 2\cos(\delta_2 - \delta_1)) \left| (\boldsymbol{\eta}_1 \cdot \boldsymbol{\epsilon}_2)^\dagger (\boldsymbol{\eta}_2 \cdot \boldsymbol{\epsilon}_1)^\dagger (\boldsymbol{\eta}_1 \cdot \boldsymbol{\epsilon}_1)(\boldsymbol{\eta}_2 \cdot \boldsymbol{\epsilon}_2) \right| \big] . \end{aligned} \qquad (5.1.8)$$

From Eqs. (5.1.7) and (5.1.8) we see explicitly that the terms of the form $(\boldsymbol{\eta}_j \cdot \boldsymbol{\epsilon}_n)$ $(j, n = 1, 2)$ determine both, the result of the intensity correlation function as well as the final state of the atomic emitters. We note that the scalar product can be expanded as

$$\boldsymbol{\eta}_j \cdot \boldsymbol{\epsilon}_n = \eta_{j_x}\epsilon_{n_x} + \eta_{j_y}\epsilon_{n_y} + \eta_{j_z}\epsilon_{n_z} = \eta_{j_+}\epsilon_{n_+} + \eta_{j_-}\epsilon_{n_-} + \eta_{j_z}\epsilon_{n_z}, \qquad (5.1.9)$$

where we used the common abbreviations $\eta_{j_\pm} := \frac{1}{\sqrt{2}}(\eta_{j_x} \pm i\, \eta_{j_y})$ and $\epsilon_{n_\pm} := \frac{1}{\sqrt{2}}(\epsilon_{n_x} \pm i\, \epsilon_{n_y})$. The basis denoted by the subscripts $+$, $-$ and z will become useful once we consider circularly polarized light.

There are different types of Λ-level structures and the polarization axes of the filters in front of the detectors can be chosen arbitrarily so that the terms $(\boldsymbol{\eta}_j \cdot \boldsymbol{\epsilon}_n)$ cannot be determined without choosing a clear reference. Nevertheless, we conclude this section by reconsidering Eq. (5.1.7): when calculating $G^{(2)}(\mathbf{r}_1, \mathbf{r}_2, \boldsymbol{\eta}_1, \boldsymbol{\eta}_2)$ we can see that it is already derived from the final projected state. In fact, being interested in the final state $|\Psi_N\rangle$ of the emitters only, we can restrict our calculation to the following ansatz

$$|\Psi_2\rangle = \hat{D}_2(\mathbf{r}_2, \boldsymbol{\eta}_2)\, \hat{D}_2(\mathbf{r}_1, \boldsymbol{\eta}_1) |e, e\rangle, \qquad (5.1.10)$$

or in general for an N-qubit state

$$|\Psi_N\rangle = \prod_{j=1}^{N} \hat{D}_N(\delta_{N-j+1}, \boldsymbol{\eta}_{N-j+1}) \prod_{n=1}^{N} |e\rangle_n. \qquad (5.1.11)$$

Let us point out that by using this definition of $|\Psi_N\rangle$ the normalization of the detector operator becomes obsolete: neither does the field amplitude $\mathcal{E}_j$ or the overall success probability $\mathcal{C}_j$ influence the normalization of the quantum state once it is projected[1], nor does the successive application of $\hat{D}_N(\delta_j, \boldsymbol{\eta}_j)$ form in general a unitary transformation. Therefore, it will be advantageous to consider the operator to be unnormalized in the following calculations and to postpone the proper normalization for the final projected quantum state.

5.1.4 Example: engineering 2-qubit quantum states

As an example, we consider a Zeeman-type Λ-level structure, labeling the Zeeman-sublevels as follows $|e\rangle := |e, m=0\rangle$, $|-\rangle := |g, m=-1\rangle$ and $|+\rangle := |g, m=+1\rangle$. The two decay channels, $|e\rangle \to |-\rangle$ and $|e\rangle \to |+\rangle$, are accompanied by the spontaneous emission of a $\boldsymbol{\epsilon}_1 = \boldsymbol{\sigma}^+$ and $\boldsymbol{\epsilon}_2 = \boldsymbol{\sigma}^-$ polarized photon, respectively. We use the common definition

---

[1] In analogy to section 2.1, the final quantum state is projected successively by means of successful measurement cycles, where the latter are depending naturally on the overall success probability $\mathcal{C}_0$ of the measurement.

of circular polarization $\sigma^+ = \mathbf{x} + i\mathbf{y}$ and $\sigma^- = \mathbf{x} - i\mathbf{y}$, where $\mathbf{x}$ ($\mathbf{y}$) denotes the linear polarization in x (y) direction.

As mentioned before, the polarization axes $\boldsymbol{\eta}_1$ and $\boldsymbol{\eta}_2$ of the filters in front of the detectors at $\mathbf{r}_1$ and $\mathbf{r}_2$, respectively, can be adjusted arbitrarily. In the following we list some particular choices of possible polarization axes $\boldsymbol{\eta}_1$ and $\boldsymbol{\eta}_2$ giving rise to some non-trivial final states of the two atomic emitters:

ϵ_1	ϵ_2	final (unnormalized) 2-qubit emitter state $|\Psi_2\rangle$
$\mathbf{x}$	$\mathbf{y}$	$(e^{i\delta_1} - e^{i\delta_2}) \cdot (|-,+\rangle - |+,-\rangle)$ $+ (e^{i\delta_1} + e^{i\delta_2}) \cdot (|+,+\rangle - |-,-\rangle)$
$\mathbf{x}+\mathbf{y}$	$\mathbf{x}-\mathbf{y}$	$i(e^{i\delta_1} - e^{i\delta_2}) \cdot (|+,-\rangle - |-,+\rangle)$ $+ (e^{i\delta_1} + e^{i\delta_2}) \cdot (|+,+\rangle + |-,-\rangle)$
σ^+	σ^-	$e^{i\delta_1} \cdot (|+,-\rangle + e^{i\delta_2}|-,+\rangle)$

From this table we see that the exact expression for the final state depends on the relative phase $\delta_{12} = \delta_2 - \delta_1$. The selection displayed in the table consists of final states that can be projected and which exhibit an interesting feature: choosing δ_1 and δ_2 appropriately, some of these projected states cannot be separated into the bare basis states of the first and the second atomic emitter, i.e., these states do not factorize in a separable emitter basis. Hence, the atomic ground states of the final state are entangled with each other. By noting that the resulting states include in particular the possibility of constructing all four Bell-states

$$\begin{aligned} |\Phi^+\rangle &= \tfrac{1}{\sqrt{2}} \cdot (|-,-\rangle + |+,+\rangle), & |\Psi^+\rangle &= \tfrac{1}{\sqrt{2}} \cdot (|-,+\rangle + |+,-\rangle), \\ |\Phi^-\rangle &= \tfrac{1}{\sqrt{2}} \cdot (|-,-\rangle - |+,+\rangle), & |\Psi^-\rangle &= \tfrac{1}{\sqrt{2}} \cdot (|-,+\rangle - |+,-\rangle), \end{aligned} \quad (5.1.12)$$

which are known to describe the complete class of maximally entangled two-qubit states (see, e.g., [101]), this fact becomes obvious.

5.2 Generation of arbitrary symmetric Dicke states in remote qubits

In the foregoing section we discussed how polarization sensitive correlation measurements of the fluorescence light scattered by few atomic emitters with a Λ-scheme can be used to engineer particular final emitter states. In the present section we are going to generalize and apply this idea for engineering specific entangled multi-partite states, namely symmetric Dicke states [21, 64].

5.2.1 Introduction to multi-partite entanglement

Multi-partite entanglement is arguably at the center of interest of most research fields related to entanglement and quantum information theory. Unfortunately, its characterization is neither fully understood nor completed and, at the moment, we only know how to classify the entanglement of a few qubits [102–104]. However, these drawbacks have not prevented the apparition of a number of proposals for generating and measuring entangled states, besides their possible applications.

The efficient and scalable preparation of entangled multi-qubit states is a key ingredient for the further characterization and experimental study of multi-partite entanglement. Several experiments have already observed genuine entangled multi-photon states [69, 71, 72, 105–110] as well as entangled distant atomic states [20, 111–113]. While some of the latter experiments are based on the exchange of photons between the qubits, there are other proposals for projecting distant non-interacting particles into entangled states via photonic measurements [12–18, 61]. Further the very recent experiments observing interference of light emitted by two atoms make use of projective measurements [37, 38, 114], which led eventually to the first demonstration of two remote entangled atoms [20].

Among the many N-qubit states, the important class of Dicke states [64] represents a particularly interesting set of quantum states associated with high robustness against particle loss [115, 116] and non-local properties of genuine entangled multi-partite states [117–119]. For example, the entangled symmetric Dicke state $|2, 0\rangle$ (see section 5.2.2) of four photonic qubits was studied in an experiment involving linear optics only [69]. In this experiment, among other features, the possibility of generating both classes of tripartite entangled states by projecting one of the four qubits was observed (see also section 5.4). Most recently, the rising interest in the class of Dicke states led to the observation of symmetric Dicke states of six photonic qubits [109, 110].

In the following, we investigate a method for generating *any* symmetric Dicke state either in distant matter or in photon polarization qubits using a multifold detection technique based on the scheme of projection introduced in section 5.1. Thereby, we grant access to the generation and measurement of an important class of genuine entangled states for potentially any number N of qubits. Our method relies on the far-field detection of N photons incoherently emitted by N initially excited atoms via spontaneous decay using suitably oriented polarizers (c.f. section 5.1). Unlike former proposals for entangling distant qubits based on projective measurements [12–18], our scheme uses explicitly the geometrical phase differences between the possible quantum paths. Furthermore, using a complementary technique, we show how to generate any symmetric Dicke state in the polarization degree of freedom of photon qubits.

5.2.2 Symmetric Dicke states of an N-qubit compound system

In quantum optics total angular momentum eigenstates are often referred to as *Dicke states* (c.f. Dicke's analyses on *coherence in spontaneous radiation process* [64]). Hereby, in the literature, the term *Dicke states* is often used in an insufficient context, referring only to the subset of symmetric Dicke states or symmetric total angular momentum eigenstates. In order to exclude any misunderstanding, in the following we use the explicit term *symmetric Dicke states* as we will analyze indeed only the symmetric subset of the Dicke states in this section 5.2. Only in the next section 5.3 we will turn our attention to the generation of arbitrary Dicke states, i.e., symmetric and non-symmetric total angular momentum eigenstates.

Let us consider an N spin-$\frac{1}{2}$ compound system with the state of the nth spin denoted by $|+\rangle_n$ or $|-\rangle_n$ if oriented up or down, respectively. The Dicke states are usually denoted by $|S_N, m_N\rangle$ and defined as the simultaneous eigenstates of both the square of the total spin operator $\hat{\mathbf{S}}^2$ and its z-component $\hat{S}_z$, where $S_N(S_N+1)\hbar^2$ and $m_N\hbar$ are the corresponding eigenvalues [120]. The $N+1$ states with the highest value of the *cooperation number* $S_N = N/2$ form a special subset of all 2^N Dicke states. These states $|\frac{N}{2}, m_N\rangle$ are the only ones which are totally symmetric under permutation of any particles and are usually written as

$$|\tfrac{N}{2}, m_N\rangle = \binom{N}{\frac{N}{2}+m}^{-\frac{1}{2}} \sum_k P_k \left(\prod_{n=1}^{\frac{N}{2}+m_N} |+\rangle_n \otimes \prod_{m=\frac{N}{2}+m_N+1}^{N} |-\rangle_m \right), \qquad (5.2.1)$$

where $\{P_k\}$ denotes the complete set of all possible distinct permutations of the qubits.

5.2.3 Description of the physical system

As introduced in section 5.1.2, our scheme considers N emitters, e.g., atoms, with a Λ-level structure where we denote the upper state by $|e\rangle$ and the two lower states by $|-\rangle$ and $|+\rangle$. The latter two ground states provide a long-lived qubit basis which we will use in the following considerations to generate quantum states utilizing the measurement scheme introduced in 5.1.3. Again, we may identify the states of the Λ-system with the Zeeman-sublevels $|e\rangle := |e, m = 0\rangle$, $|-\rangle := |g, m = -1\rangle$ and $|+\rangle := |g, m = +1\rangle$, where the excited state $|e\rangle$ has two decay channels, $|e\rangle \rightarrow |-\rangle$ and $|e\rangle \rightarrow |+\rangle$, accompanied by the spontaneous emission of a $\boldsymbol{\sigma}^+$ ($\boldsymbol{\sigma}^-$) polarized photon. As described in section 5.1, for a single atom with a Λ-level structure, the polarization state of the emitted photon is entangled with the corresponding ground state of the de-excited atom [44, 100]. The total state of the atom and the photon before a detection of the photon is performed can be described by (c.f. Eq. (5.1.2))

$$|\Phi\rangle = \frac{1}{\sqrt{2}}(|-\rangle|\boldsymbol{\sigma}^+\rangle + |+\rangle|\boldsymbol{\sigma}^-\rangle). \tag{5.2.2}$$

Let us now turn to the setup introduced in section 5.1.2. The alignment of the N atoms and the jth detector equipped with an individual polarization analyzer is depicted in figure 5.1. In the case of $N = 1$, after a detector has recorded the emitted photon with a polarization equal to $\boldsymbol{\sigma}^+$ ($=\boldsymbol{\sigma}^-$), the atom is projected into the ground state $|-\rangle$ ($|+\rangle$). However, for $N > 1$, the detectors located in the far-field region of the atoms are unable to distinguish which particular atom has emitted a registered photon. Therefore, after the detection of a first photon, all atoms will form a correlated state [12–18].

The entanglement of the atoms is a consequence of two ingredients: the impossibility of the detectors to determine which atom emitted a particular photon together with the projection postulate which states that after the detection of a photon the state of the atoms is projected into a state compatible with the outcome of the measurement [12]. In the following we introduce a third ingredient to this scheme which is well-known from chapter 4: the idea to exploit the geometrical phase differences of the $N!$ quantum paths resulting from the $N!$ possibilities that each of the N atoms emits a photon which is subsequently registered by one of the N detectors. As will be shown below, these geometrical phase differences will allow us to prepare symmetric Dicke states of arbitrary configuration.

In the following, we restrict the orientation of the polarizer axes in our setup to $\boldsymbol{\eta}_j = \{\boldsymbol{\sigma}^+, \boldsymbol{\sigma}^-\}$, i.e., sensitive solely to photons polarized along $\boldsymbol{\sigma}^+$ or $\boldsymbol{\sigma}^-$. The (unnormalized) operator describing the detection event of the jth photon at $\mathbf{r}_j$ can thus be written in the form

$$\hat{D}_N(\delta_j, \boldsymbol{\eta}_j) = \sum_{n=1}^{N} e^{in\delta_j} |x_j\rangle_n \langle e|, \tag{5.2.3}$$

where δ_j is the phase introduced in Eq. (2.1.2) and the operator $|x_j\rangle_n \langle e|$ projects the nth atom from state $|e\rangle$ to the ground-state $|x_j\rangle \in \{|-\rangle, |+\rangle\}$. The final projected state $|x_j\rangle$ depends uniquely on the adjustment of the polarizer in front of the jth detector, which projects the polarization of the measured photon along $\boldsymbol{\eta}_j$. Please note that Eq. (5.2.3) is directly derived from Eq. (5.1.4) for $\boldsymbol{\eta}_j = \{\boldsymbol{\sigma}^+, \boldsymbol{\sigma}^-\}$, $\boldsymbol{\epsilon}_1 = \boldsymbol{\sigma}^+$, $\boldsymbol{\epsilon}_2 = \boldsymbol{\sigma}^-$, $|g_1\rangle = |-\rangle$ and $|g_2\rangle = |+\rangle$.

5.2.4 Preparation of symmetric 3-qubit Dicke states

With the detection operator of Eq. (5.2.3) we can describe the detection processes of all N photons emitted by the N atoms. As an example, let us consider the case of $N = 3$ qubits. After a first photon is detected at $\mathbf{r}_1$, we obtain from Eqs. (5.1.3) and (5.2.3):

$$\hat{D}_3(\delta_1, \boldsymbol{\eta}_1) |\Psi_3^i\rangle = e^{i\delta_1} |x_1, e, e\rangle + e^{i2\delta_1} |e, x_1, e\rangle + e^{i3\delta_1} |e, e, x_1\rangle. \qquad (5.2.4)$$

The detection of the second and third photon may occur at $\mathbf{r}_2$ and $\mathbf{r}_3$ and we describe these events by applying successively the two detector operators $\hat{D}_3(\delta_2, \boldsymbol{\eta}_2)$ and $\hat{D}_3(\delta_3, \boldsymbol{\eta}_3)$ on the intermediate state $\hat{D}_3(\delta_1, \boldsymbol{\eta}_1) |\Psi_3^i\rangle$. The final state $|\Psi_3^f\rangle$ of the three atoms can then be written as:

$$\begin{aligned} |\Psi_3^f\rangle :=\ & \hat{D}_3(\delta_3, \boldsymbol{\eta}_3) \hat{D}_3(\delta_2, \boldsymbol{\eta}_2) \hat{D}_3(\delta_1, \boldsymbol{\eta}_1) |\Psi_3^i\rangle \\ =\ & e^{i\delta_1+i2\delta_2+i3\delta_3} |x_1, x_2, x_3\rangle + e^{i\delta_1+i2\delta_3+i3\delta_2} |x_1, x_3, x_2\rangle \\ +\ & e^{i\delta_2+i2\delta_1+i3\delta_3} |x_2, x_1, x_3\rangle + e^{i\delta_3+i2\delta_1+i3\delta_2} |x_3, x_1, x_2\rangle \\ +\ & e^{i\delta_2+i2\delta_3+i3\delta_1} |x_2, x_3, x_1\rangle + e^{i\delta_3+i2\delta_2+i3\delta_1} |x_3, x_2, x_1\rangle. \end{aligned} \qquad (5.2.5)$$

For three equidistantly aligned atoms and polarizers oriented along $\boldsymbol{\eta}_j = \{\boldsymbol{\sigma}^+, \boldsymbol{\sigma}^-\}$ this is the most general expression of the final state. As can be seen from Eq. (5.2.5), the geometrical phase differences δ_j, $j = 1, \ldots, 3$, determine the symmetry of the state. In particular, to generate symmetric Dicke states, the phases δ_j should adopt multiple values of 2π, which can be achieved by a suitable localization of the N detectors according to Eq. (2.1.2). Note that the final form of the state (5.2.5) depends eventually on the orientation of the polarization analyzers in front of the detectors: If the jth polarizer is oriented to transmit $\boldsymbol{\sigma}^+$ ($\boldsymbol{\sigma}^-$) polarized light, the internal levels of the atoms will be projected onto the state $|x_j\rangle = |-\rangle$ ($|x_j\rangle = |+\rangle$). In particular, this means that for $\delta_j = n\, 2\pi$ with n being an integer ($j = 1, 2, 3$) we can generate all four symmetric three-qubit Dicke states

$$\begin{aligned} |\tfrac{3}{2}, +\tfrac{3}{2}\rangle &= |+, +, +\rangle \\ |\tfrac{3}{2}, +\tfrac{1}{2}\rangle &= \tfrac{1}{\sqrt{3}} (|+, +, -\rangle + |+, -, +\rangle + |-, +, +\rangle) \\ |\tfrac{3}{2}, -\tfrac{1}{2}\rangle &= \tfrac{1}{\sqrt{3}} (|+, -, -\rangle + |-, +, -\rangle + |-, -, +\rangle) \\ |\tfrac{3}{2}, -\tfrac{3}{2}\rangle &= |-, -, -\rangle. \end{aligned} \qquad (5.2.6)$$

The simple product state $|\frac{3}{2}, +\frac{3}{2}\rangle$ ($|\frac{3}{2}, -\frac{3}{2}\rangle$) can be obtained by orienting the three polarizers to transmit σ^- (σ^+) polarized light only so that all atoms are projected onto the state $|+\rangle$ ($|-\rangle$). It is, however, also possible to generate the genuine tripartite entangled state $|\frac{3}{2}, +\frac{1}{2}\rangle$ ($|\frac{3}{2}, -\frac{1}{2}\rangle$). In this case, one polarizer should be oriented to transmit σ^+ (σ^-) polarized and two polarizers to transmit σ^- (σ^+) polarized light. Hereby it does not matter which of the three detectors is actually sensitive to σ^+- or σ^- polarized photons, since all detectors are placed in the far-field region of the atoms and cannot distinguish individual photon sources.

5.2.5 Preparation of symmetric N-qubit Dicke states

So far we showed how to generate all the symmetric Dicke states for $N = 3$ atoms. The generalization to an arbitrary number N of atoms is nevertheless straightforward. For this, we have to place again all N detectors at positions $\mathbf{r}_1, \ldots, \mathbf{r}_N$ such that the phases δ_j, $j = 1, \ldots, N$, adopt multiple values of 2π. The state of the N atoms after a first photon has been detected at $\mathbf{r}_1$ can be calculated by applying the operator $\hat{D}_N(\delta_1, \boldsymbol{\eta}_1)$ on the initial state (5.1.3). From this we obtain

$$\hat{D}_N(\delta_1, \boldsymbol{\eta}_1)|\Psi_N^i\rangle = \sum_k P_k(|x_1, e, \ldots, e\rangle_N), \tag{5.2.7}$$

where $\{P_k\}$ denotes the set of all possible permutations of the N qubits.

In analogy to the case $N = 3$, we assume that the $N - 1$ remaining photons are detected at positions $\mathbf{r}_2, \mathbf{r}_3, \ldots, \mathbf{r}_N$, respectively. We can calculate the final state of the atoms, after all N photons have been detected at the N detectors, by applying the $N - 1$ detector operators $\hat{D}_N(\delta_2, \boldsymbol{\eta}_2)$, $\hat{D}_N(\delta_3, \boldsymbol{\eta}_3)$, ..., $\hat{D}_N(\delta_N, \boldsymbol{\eta}_N)$ on the intermediate state (5.2.7). From this we obtain:

$$|\Psi_N^f\rangle = \sum_k P_k(|x_1, x_2, \ldots, x_N\rangle_N). \tag{5.2.8}$$

With the final state of the N atoms given by Eq. (5.2.8), we still have to choose the orientation of the N polarizers to determine the final state of the N qubits $|x_j\rangle$ ($j = 1, \ldots, N$). In particular, if we want to generate the symmetric Dicke state $|\frac{N}{2}, m_N\rangle$, with $m_N \in -\frac{N}{2}, \ldots, \frac{N}{2}$, we have to choose $\frac{N}{2} + m_N$ polarizers to be sensitive to σ^- polarized light and $\frac{N}{2} - m_N$ polarizers to be sensitive to σ^+ polarized light; this will determine the final state of the atoms to contain $\frac{N}{2} + m_N$ qubits in the state $|+\rangle$ and $\frac{N}{2} - m$ in the state $|-\rangle$. Again assuming that each detector registers one and only one photon, the atoms are projected into the state $|\frac{N}{2}, m_N\rangle$ which - for arbitrary m_N and N - represents any symmetric Dicke state (c.f. Eq. (5.2.1)).

5.2.6 Entanglement at remote distances by using a detection scheme based on optical fibers

In principle, our method does not require nearby particles since we do not make use of any interaction between the atoms. Nevertheless the far-field condition inherent in expression (2.1.2), and therefore in Eqs. (5.2.3)-(5.2.8), implies a practical limit for the separation between the particles. However, as described in section 3.5, this limit can be overcome by the use of optical fibers. Connecting each of the N atoms with all N detectors by using N^2 identical fibers leads to the $N!$ possible quantum paths as well. Hereby, the optical phases are no longer determined by the condition (2.1.2) but simply by the optical paths between each ion and the light collecting detectors. Placing the N^2 identical fibers such that all collected photons travel the same distance from the ions to the detectors, the condition $\delta_j = 2\pi$ is again fulfilled. Note that optical fibers are commonly used in experiments involving single atoms to collect the light of selective modes, see, e.g., [38,44]. In this way we can apply our scheme even to spatially far distant, i.e., *remote*, particles.

The idea of substituting the optical free space pathways from the N atoms to one detector by N optical fibers will be discussed in more detail in section 5.3. Apart from achieving the loss of which-way information, where the detectors cannot *distinguish* the source of the registered photons, optical fibers moreover enable us to utilize a greater parameter space than is present for a far-field based detection scheme: a photon traveling along any of N optical fibers, i.e., any of the N single optical pathways, may accumulate a unique, distinct optical phase different from all other $N-1$ pathways. In section 5.3, we will see that this greater parameter space can be fruitfully exploited and used to generate more complex classes of multi-partite entangled quantum states [24].

5.2.7 Experimental feasibility

Let us take a look at the experimental feasibility of our proposed scheme to generate arbitrary symmetric Dicke states in a measurement based projection scheme. In section 4.2.4, we already addressed the technical feasibility to achieve a phase resolution suitable for the purpose of quantum imaging. Thereby, our attention focused on the resolution of the Nth order correlation signal and its visibility. Now, for the purpose of quantum state engineering, we will take a look at the counting rate of our measurement and the fidelity of the final projected state, i.e., the *closeness* of the desired state and the actual state being generated.

In an experimental realization of the far-field detection scheme there is a natural trade-off between the counting rate of the scattered photons and the fidelity of the final state: the former increases with the size of the detector, while the latter decreases with it. Further, the counting rate profits from a high repetition rate, a large detection area being covered,

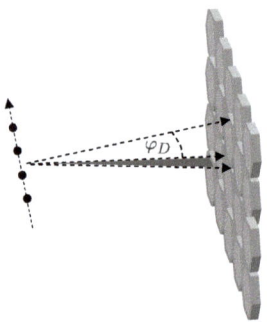 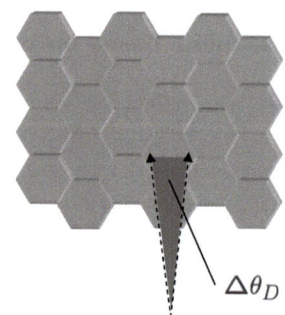

Figure 5.2: Optimized setup for the generation of a four-partite symmetric Dicke states used for the estimation of its technical feasibility. Using the fact that the setup provides perfect symmetry along φ_D we can maximize the counting rate by using an extended CCD-camera to cover large angles φ_D. Further, by limiting the azimuthal detection windows to a small angles $\Delta\theta_D$, the influence of an uncertainty in the trapping confinement of the atoms along the trap axis can be neglected.

e.g., by using CCD-cameras, and a high quantum efficiency of the detectors, whereas a fidelity close to unity requires a high degree of localization of the atomic emitters.

The uncertainty in the final projected state is determined mainly by the uncertainty within the optical phases $\Delta\delta_j$[2]. The latter is determined by two geometrical measures: the finite solid angle covered by the detector surface and the confinement of the atoms. For our estimate we consider ions localized in a linear trap and a detector arrangement as depicted in figure 5.2. Using a worst case estimate, the maximal possible error in phase is given by

$$\Delta\delta_{max} \approx k \left(\Delta d_{vert.} + \left(\frac{1}{2} - \frac{\cos \Delta\theta_D}{2} \right) d \right), \qquad (5.2.9)$$

where $\Delta d_{vert.}$ is the uncertainty in the atomic confinement in lateral direction, i.e., perpendicular to the trap axis, and $\Delta\theta_D$ is the azimuthal detection window as determined by the detector surface (c.f. figure 5.2). We note that uncertainties within the atomic confinement in direction of the atomic alignment[3] and within the wavelength k were omitted since they do not play a major role.

With $\Delta\delta_{max}$ determined by Eq. (5.2.9), we can estimate the expected fidelity of the final state projected by our setup, e.g., for generating the symmetric Dicke state $|2,0\rangle$ using

---

[2] We note that uncertainties within the polarization filters can be suppressed below 10^{-10} and thus do not play any role in our considerations [121].

[3] The uncertainty within the atomic confinement along the alignment does not contribute to $\Delta\delta_{max}$ if the detection plane is close to a plane perpendicular to the alignment (c.f. figure 5.2).

$N = 4$ adjacent atoms. We assume the atoms to be 5 µm apart from each other and confined to $\Delta d_{vert.} = 5$ nm. Further, the azimuthal detection window is $\Delta \theta_D = 2°$ and the wavenumber used $k = 2\pi/800$ nm. Including all these uncertainties in our analysis via error propagation, we estimate $\Delta \delta_{max.} \approx 0.05$ and a fidelity of 90% for the generation of the four qubits state $|2, 0\rangle$. Remarkably it was shown recently that a fidelity of 66% is already sufficient to demonstrate the entanglement of this state [117].

The overall success probability C_0 to detect a single scattered photon is proportional to the solid angle $\Delta\Omega$ covered by the detector surface divided by 4π (c.f. Eq. (3.1.2)). With r being the repetition rate of our experiment, the counting rate R_N for detecting N photons is then given by $R_N = r\, C_0^N = r\, (\mu \frac{\Delta\Omega}{4\pi})^N$. The setup shown in figure 5.2 depicts a detection scheme which not only allows to optimize $\Delta\delta_{max}$ but also the counting rate: by adjusting a large number of detectors in vertical direction, e.g., by using CCD-cameras, one can increase the total detection surface leading to successful measurements for a given $\Delta\theta_D$. Therefore, the counting depends on the angle $\Delta\varphi_D = max(\varphi_D) - min(\varphi_D)$, on the azimuthal detection window, the quantum efficiency of the detectors and the repetition rate. In an experiment that uses CCD-cameras covering a fair area in the detection plane, e.g., $\Delta\theta_D = 2°$ and $\Delta\varphi_D = 60°$, and taking into account all sources of errors mentioned above, we expect a counting rate $R_4 = r\, (\mu \frac{\Delta\theta_D}{6\pi})^4$ of the needed four-fold coincident events to be of the order of a few mHz with an excitation rate of several tens of MHz [100] and a quantum efficiency of about 50% (see [19]). In general the counting rate decreases with the number of qubits. This might limit the scalability of our scheme as is indeed the case with other experiments observing entangled photons [69, 105–110] as well as entangled atoms [20, 44, 100].

5.2.8 Generation of symmetric Dicke states in photon qubits

Finally, we want to discuss how our method can also be used to prepare symmetric Dicke states in the polarization degree of freedom of photon qubits. Recently the Dicke states $|2, 0\rangle$ and $|3, 0\rangle$ have been observed as entangled photon polarization states in a postselective manner, by using initially entangled photons generated in SPDC [69, 109, 110]. To prepare arbitrary symmetric Dicke states of photon polarization qubits we have to place the polarization analyzers, formerly positioned in front of the detectors (see figure 5.1), in front of the atoms such that the polarization of each spontaneously emitted photon is determined by an individual polarizer. The setup remains otherwise identical to the one presented above: all N atoms are initially prepared in the excited state $|e\rangle$ and, via postselection, we assure that one and only one photon is registered at each of the N detectors. However, after the detection of the photons the internal state of each atom is now uniquely determined by the orientation of the polarizer, i.e., in correspondence to the polarization state of the photon emitted by this particular atom. Since the photons are still detected in the far-field region of the atoms, we do not acquire any which-way information of individual photons and thus cannot determine the polarization state of each individual photon at any of the N detectors. Instead, all N quantum paths associated with the N possibilities that a photon has been emitted by one of the N atoms will contribute to a single-photon detection event at a particular detector.

Introducing the wave vectors of the N different spatial modes $\mathbf{k}_j = k\,\mathbf{e}_j$, defined by the unit vector $\mathbf{e}_j := \mathbf{r}_j/|\mathbf{r}_j|$ pointing in the direction of the jth detector, we only know after the detection of all N photons at $\mathbf{r}_1, \ldots, \mathbf{r}_N$ that each single mode $\mathbf{k}_j$ was populated by exactly one photon. But what was the polarization state of the photon in the jth mode? In this context, it is suggestive to denote the polarization state of a photon in the mode $\mathbf{k}_j$ as $|k_j\rangle = \{|\sigma^+\rangle, |\sigma^-\rangle\}$ instead of $|x_j\rangle = \{|+\rangle, |-\rangle\}$. Keeping the detector positions as for generating Dicke states of massive particles, we obtain the same state as given in Eq. (5.2.8), however now for the polarization state of the N photons in the N spatial modes. It is thus possible to generate an arbitrary symmetric Dicke state $|\frac{N}{2}, m_N\rangle$ of photon polarization qubits, by choosing $\frac{N}{2} + m_N$ polarizers to transmit σ^- and $\frac{N}{2} - m_N$ polarizers to transmit σ^+ polarized light only.

5.2.9 Conclusions

In section 5.2, we demonstrated that it is possible to generate all symmetric Dicke states for distant matter as well as for photon polarization qubits using linear optical tools only [21]. Our method offers thus a simple access to genuine entangled states of any number of qubits exploiting the absence of which-way information and polarization sensitive far-field detection of photons spontaneously emitted by atoms in a Λ-configuration based on a scheme of projection as introduced in section 5.1. As for the technical feasibility of using optical phase differences between single ions, we refer again to [94] where first order interferences of light coherently scattered by two ions were observed.

It can be seen from Eq. (5.2.5) that our method is also capable of generating entangled quantum states different from the symmetric Dicke states: utilizing the degrees of freedom offered by the use of optical fibers or utilizing the arbitrary orientations of the polarizers it is possible to generate other classes of entangled quantum states in long-lived matter qubits [22, 24]. This will be investigated in the following two sections in more detail.

5.3 Generation of arbitrary total angular momentum eigenstates

In this section we will generalize the results found in section 5.2, however, by using novel degrees of freedom. In particular, we will make use of the individual optical path lengths between each atom and detector offered by the use of optical fibers. This can be fruitfully exploited in order to generate arbitrary total angular momentum eigenstates, i.e., symmetric and non-symmetric Dicke states [24].

5.3.1 Introduction: coupling of angular momenta of non-interacting qubits

Since the celebrated article by Einstein, Podolsky, and Rosen in 1935 [1], it is commonly assumed that the phenomenon of entanglement between different quantum systems occurs if the systems *had previously interacted with each other*. Indeed, for most experiments generating entangled quantum states, interactions such as non-linear effects [5], atomic collisions [6,7], Coulomb coupling [8,9], or atom-photon interfaces [10], are a prerequisite. However, as mentioned before, recent proposals considered that entanglement between systems that never interacted before can be created as a consequence of measuring photons propagating along multiple quantum paths, leaving the emitters in particular entangled states [12–18,21,22,61]. So far, several experiments generating *entanglement at a distance via projection* have been realized, first between disordered clouds of atoms [111–113] and very recently even between single trapped atoms [20].

On the other hand, the coupling of angular momenta is commonly utilized to account for the interaction between particles in order to retrieve the corresponding energy eigenstates and eigenvalues of the total system. The coupling of angular momenta has been fruitfully employed in as disparate fields as solid state, atomic or high-energy physics, to account for the interaction between electric or magnetic multipoles or spins of quarks, respectively [122]. Here again, it seems counter-intuitive that non-interacting particles, such as remotely placed spin-1/2 particles, couple to form arbitrary total angular momentum eigenstates including highly and weakly entangled quantum states as if an interaction were present.

In the following, we describe a method how to *mimic* the universal coupling of the angular momenta of N remote non-interacting spin-1/2 particles (qubits) in an experimentally operational manner. Hereby, an arbitrary number of distant particles can be entangled in their two-level ground states providing long-lived qubit states via the use of suitably designed projective measurements. In reference to the algorithm describing the coupling of angular momenta of individual spin-1/2 particles, our method couples successively remote qubit states to a multi-qubit compound system. Thereby, it offers access to the entire coupled basis of an N-qubit compound system of dimension 2^N, i.e., to any of the 2^N symmetric and non-symmetric total angular momentum eigenstates.

5.3.2 Description of the physical system

In the foregoing section 5.2 we already gave attention to the generation of a subset of total angular momentum eigenstates, namely the symmetric states Dicke states. In this section, however, we intend to investigate the generation of a broader class of states, i.e., the symmetric and non-symmetric total angular momentum eigenstates. For this purpose, let us briefly recall the general definition of angular momentum eigenstates (c.f. section 5.2.2): for N spin-1/2 particles, the total angular momentum eigenstates, defined as simultaneous eigenstates of the square of the total spin operator $\hat{S}^2$ and its z-component $\hat{S}_z$, are commonly denoted by $|S_N;m_N\rangle$, with the corresponding eigenvalues $S_N(S_N+1)\hbar^2$ and $m_N\hbar$ (see section 5.2.2) [64, 120]. However, we note that there are generally several distinct ways how N elementary spin-1/2 particles can couple to a total spin of S_N. Although the notation is unambiguous when considering the subset of symmetric states only, i.e., where $S_N = N/2$ (c.f. section 5.2). Since the denomination $|S_N;m_N\rangle$ characterizes thus generally more than one quantum state, we will extend the notation of an N-qubit state by its coupling history, i.e., by adding the values of $S_1, S_2, ..., S_{N-1}$ to those of S_N and m_N. A single qubit state has $S_1 = \frac{1}{2}$, a two-qubit system can either have $S_2 = 0$ or $S_2 = 1$, a three-qubit system $S_3 = \frac{1}{2}$ or $S_3 = \frac{3}{2}$, and so on. Including the coupling history, i.e., the values of $S_1, S_2, ..., S_{N-1}$, we thus get the following notation $|S_1, S_2, ..., S_N;m_N\rangle$ which describes a particular angular momentum eigenstate unambiguously.

Again, we consider a system consisting of N labeled but indistinguishable single-photon emitters with a Λ-configuration as shown in figure 5.3. As before, it is hereby most reasonable to consider trapped ions due to the high extend to which they can be controlled. We denote the two ground levels of the Λ-configured atoms as $|+\rangle$ and $|-\rangle$ or, using the notation introduced before, as $|\frac{1}{2};+\frac{1}{2}\rangle \equiv |+\rangle$ and $|\frac{1}{2};-\frac{1}{2}\rangle \equiv |-\rangle$. Initially, all atoms are excited by a laser π pulse towards the excited state $|\Psi_N^i\rangle$ (c.f. Eq. (5.1.3)) and subsequently decay by spontaneously emitting N photons.

In contrast to the basic setup of section 5.2 as depicted in figure 5.1, we now consider a setup where the photons are collected by single-mode optical fibers [20,44] and transmitted to the N different detectors. Since each atom is connected via optical fibers to several detectors, a single photon can travel on several alternative, yet equally probable pathways to be eventually recorded by one detector. After a successful measurement, where all N photons have been recorded at the N detectors so that each detector registers exactly one photon, it is thus impossible to determine along *which way* each of the N photons propagated - in analogy to a far-field based measurement scheme discussed in sections 3.5

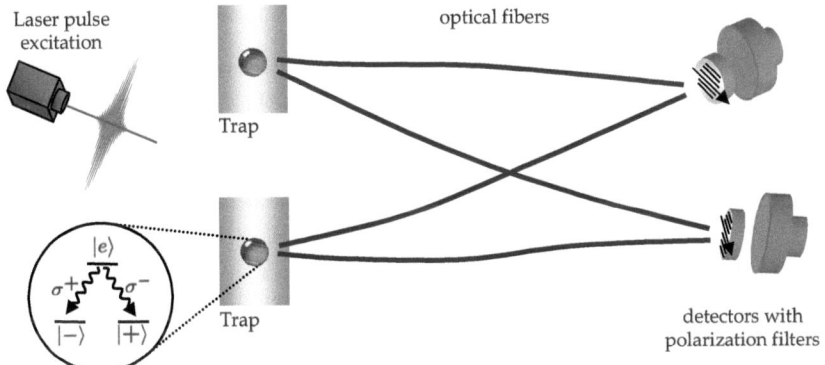

Figure 5.3: Experimental setup for the angular momentum coupling of two atoms via projective measurements using optical fibers. In a successful measurement cycle, each atom emits a single photon and each detector registers a single photon. Note that the detector cannot distinguish which of the atoms emitted a registered photon.

and 5.2.6. This may cause quantum interferences of Nth order which can be fruitfully employed to engineer particular quantum states of the emitters, e.g., to generate families of entangled states symmetric under permutation of their qubits [21, 22]. In the following, we will demonstrate how to generate any quantum state belonging to the coupled basis of an N-qubit compound system by mimicking the process of spin-spin coupling, including symmetric *and* non-symmetric states.

5.3.3 Preparation of total angular momentum eigenstates

Let us start by recalling the most basic process of our system (c.f. section 5.1): if one single excited atom with a Λ-level structure emits a photon, the atomic ground state and the photonic polarization states cannot be described independently. The excited state $|e\rangle$ can decay along two possible channels, $|e\rangle \to |+\rangle$ and $|e\rangle \to |-\rangle$, accompanied by the spontaneous emission of a σ^- or a σ^+ polarized photon, respectively, e.g., if Zeeman sub-levels are considered. A single decaying atom thus forms an entangled state between the polarization state of the emitted photon and the corresponding ground state of the de-excited atom [44, 100] (c.f. Eqs. (5.1.2) and (5.2.2)). This correlation implies that the state of the atom is projected onto $|+\rangle$ ($|-\rangle$) if the emitted photon is registered by a polarization sensitive detector transmitting only σ^- (σ^+) polarized light.

Preparation of 2-qubit states

In a next step, we consider the system shown in figure 5.3 where two atoms with Λ-level structure are initially excited and subsequent measurements on the spontaneously emitted photons are performed at two different detectors. Again, implementing a polarization sensitive measurement on the two emitted photons using two different polarization filters in front of the detectors, the state of the two atoms is projected due to the measurement. However, if the polarization of both photons is measured along orthogonal directions, the state of the atoms will be projected onto a superposition of both ground states, since it is impossible to determine which atom emitted the photon traveling to the first or to the second detector by the information obtained in the measurement process.

With each qubit having a total spin of $\frac{1}{2}$, a two-qubit system can have a total spin of either 1 or 0 and thus defines four angular momentum eigenstates given by:

spin-1 triplet	$\|S_1,S_2;m_2\rangle$	**spin-0 singlet**	$\|S_1,S_2;m_2\rangle$
$\|++\rangle$	$\|\frac{1}{2},1;+1\rangle$		
$\frac{1}{\sqrt{2}}(\|+-\rangle+\|-+\rangle)$	$\|\frac{1}{2},1;0\rangle$	$\frac{1}{\sqrt{2}}(\|+-\rangle-\|-+\rangle)$	$\|\frac{1}{2},0;0\rangle$
$\|--\rangle$	$\|\frac{1}{2},1;-1\rangle$		

The spin-1 triplet can be easily generated with the setup shown in figure 5.3 and by choosing the polarization filters accordingly: for example, if both filters are oriented in such a way that only σ^- (σ^+) polarized photons are transmitted, the emitters are projected onto the state $|++\rangle$ ($|--\rangle$); if the filters are orthogonal, i.e., one is transmitting σ^- and one σ^+ polarized photons, the system is projected onto the state $|\frac{1}{2},1;0\rangle$, since any information along *which way* the photons propagated is erased by the system. Finally, in order to generate the singlet state $|\frac{1}{2},0;0\rangle$, we may introduce an optical phase shift of π in one of the optical paths shown in figure 5.3, e.g., by extending or shortening the length of the optical path by $\frac{\lambda}{2}$. The generation of the four two-particle total angular momentum eigenstates with the system shown in figure 5.3 thus requires only the variation of two polarizer orientations and, in case of the singlet state, to introduce an optical phase shift of π.

Preparation of 3-qubit states

With the two-qubit total angular momentum eigenstates at hand, we can next couple an additional qubit in order to access the eight possible three-qubit total angular momentum eigenstates. In the following, we will exemplify our method for the three-qubit state

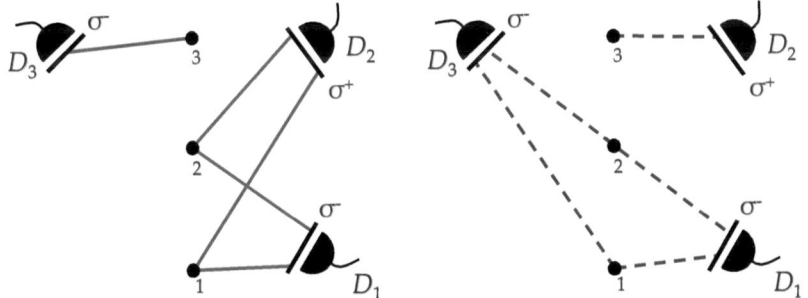

Figure 5.4: Left: Extension of the setup shown in figure 5.3 capable of generating the state $|\frac{1}{2},1;0\rangle \otimes |+\rangle$. Right: configuration for the generation of $|\frac{1}{2},1;1\rangle \otimes |-\rangle$. The optical fibers connecting the detectors with the emitters are symbolized by red solid (left) and blue dashed (right) lines, respectively. All fibers have to be identical (up to phase differences of multiples of 2π).

$|\frac{1}{2},1,\frac{1}{2};+\frac{1}{2}\rangle$ given by

$$|\tfrac{1}{2},1,\tfrac{1}{2};+\tfrac{1}{2}\rangle = \frac{1}{\sqrt{6}}\left(2|++-\rangle - |+-+\rangle - |-++\rangle\right) \qquad (5.3.1)$$
$$= \sqrt{\frac{2}{3}}|\tfrac{1}{2},1;+1\rangle \otimes |-\rangle - \frac{1}{\sqrt{3}}|\tfrac{1}{2},1;0\rangle \otimes |+\rangle,$$

where the last line in Eq. (5.3.1) exhibits the coupling history: in order to generate the three-qubit state $|\frac{1}{2},1,\frac{1}{2};+\frac{1}{2}\rangle$, the two-qubit spin-1 states $|\frac{1}{2},1;+1\rangle$ and $|\frac{1}{2},1;0\rangle$ are coupled with $|-\rangle$ and $|+\rangle$, respectively. Thereby, the prefactors $\sqrt{\frac{2}{3}}$ and $-\frac{1}{\sqrt{3}}$ represent the corresponding Clebsch-Gordan coefficients associated with the change of the basis [123]. In the following, we will make use of our knowledge of how to obtain the states $|\frac{1}{2},1;+1\rangle$ and $|\frac{1}{2},1;0\rangle$ in order to generate the desired state $|\frac{1}{2},1,\frac{1}{2};+\frac{1}{2}\rangle$. Therefore, we have to add a third qubit and combine the two setups generating the two individual states accordingly in one setup.

The two setups individually capable of generating the three-qubit states $|\frac{1}{2},1;+1\rangle \otimes |-\rangle$ and $|\frac{1}{2},1;0\rangle \otimes |+\rangle$ are shown in figure 5.4. The additional qubit is not yet coupled to the two-qubit system, i.e., it is simply projected either onto the state $|+\rangle$ (figure 5.4, left) or $|-\rangle$ (figure 5.4, right), whereas the two-qubit systems are projected in the same way as explained above in the context of figure 5.3. In order to generate the three-qubit state $|\frac{1}{2},1,\frac{1}{2};\frac{1}{2}\rangle$, we now have to superpose these two possibilities. The combined system is shown in figure 5.5. We will explain the underlying physics by considering the possible scenarios when detecting the photon emitted by the additional third atom.

In a successful measurement cycle, the three emitted photons are detected at three dif-

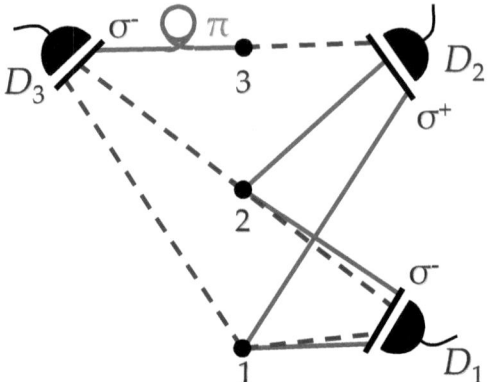

Figure 5.5: Setup for the generation of the state $2|++-\rangle - |+-+\rangle - |-++\rangle$. The red solid lines indicate the path which leads to $-|+-+\rangle - |-++\rangle$, whereas the blue dashed labeled path leads to $2|++-\rangle$. The minus sign is due to the phase shift of π introduced between atom 3 and detector D_3. *Note*: The different red solid and blue dashed lines leading from atom 1 (2) to detector D_1 are drawn to indicate the different optical paths only. Physically, there is only one fiber from atom 1 (2) to detector D_1.

ferent detectors. Thus, there are only two possible situations due to a measurement of a photon emitted by the third atom:

I. (red solid lines) The emitted photon is registered at detector D_3 which has a σ^- polarizing filter in front. In this case, emitter 3 is projected onto the state $|+\rangle$ and emitter 1 and 2 are left in the setup generating the state $|\frac{1}{2},1;0\rangle \equiv \frac{1}{\sqrt{2}}(|+-\rangle + |-+\rangle)$; as discussed in figure 5.4, left.

II. (blue dashed lines) The emitted photon is registered at detector D_2 which has a σ^+ polarizing filter in front. In this case, emitter 3 is projected onto the state $|-\rangle$ and emitter 1 and 2 are left in the setup generating the state $|\frac{1}{2},1;1\rangle \equiv |++\rangle$; as discussed in figure 5.4, right.

In other words, the third emitter acts as a switch between the two possible quantum paths: with equal probabilities, the system is either projected onto the state $2|++-\rangle$ or onto the state $|+-+\rangle + |-++\rangle$. Note that the relative factor of two results from using an equal number of pathways (optical fibers) in both cases. In addition, we can modify the path where a photon emitted by the third atom is registered at detector D_3 by implementing a relative optical phase shift of π (c.f. figure 5.5) to obtain a minus sign for scenario II. relative to scenario I. In this case, the final state projected by the setup shown in figure 5.5 corresponds to the three-qubit state $|\frac{1}{2},1,\frac{1}{2};\frac{1}{2}\rangle$ of Eq. (5.3.1) as announced.

Reconsidering the state $|\frac{1}{2},1,\frac{1}{2};\frac{1}{2}\rangle$ in terms of our extended notation, we coupled two spin-1/2 particles to form a spin-1 compound state that was coupled again with a spin-1/2 particle to form a three-particle spin-1/2 compound state. Similarly, we could have coupled the spin-1 compound state with an additional qubit in such a way that we obtain the symmetric state $|\frac{1}{2},1,\frac{3}{2};\frac{1}{2}\rangle$, also known as W-state [102]. For this case, we have to modify the setup shown in figure 5.5 slightly: we remove the optical phase shift of π and connect the third emitter also with detector D_1. In this case, the totally symmetric setup generates a W-state as already discussed and outlined in section 5.2 [21].

Preparation of N-qubit states

Finally, let us outline how to engineer the coupling of angular momenta of N remote qubits to form an arbitrary N-qubit total angular momentum eigenstate. In order to generate the N-qubit state $|S_1, S_2, S_3, ...S_N; m_N\rangle$ we have to

1. set up $\frac{N}{2} + m_N$ ($\frac{N}{2} - m_N$) detectors with polarization filters in front transmitting σ^- (σ^+) polarized photons. Furthermore, we connect the first emitter with optical fibers to all N detectors.

2. check for each additional qubit i beginning with $i = 2$ whether $S_i > S_{i-1}$ or $S_i < S_{i-1}$. If

    a. $S_i < S_{i-1}$; we have to connect the emitter i with optical fibers to one detector with a σ^- polarizer and to one with a σ^+ polarizer. The optical fiber leading to the σ^- polarizer should induce a relative optical phase shift of π and these two detectors should not be linked to any other subsequent emitter j ($j > i$).

    b. $S_i > S_{i-1}$; we have to connect the emitter i with optical fibers to all detectors except those which are excluded by previous applications ($j < i$) of case a.

If one wants to create a particular total angular momentum eigenstate of the form $|S_1, S_2, S_3, ...S_N; m_N\rangle$, the setup is determined by the total spins $S_1, S_2, S_3, ...S_N$ obtained by successively coupling N spin-1/2 particles. Hereby, the spin number m_N determines the fraction of polarization filters transmitting σ^- and σ^+ polarized light which are used in the setup (see figure 5.6).

Let us apply this algorithm for the example of the three-qubit total angular momentum eigenstates $|\frac{1}{2},1,\frac{1}{2};\frac{1}{2}\rangle$ and $|\frac{1}{2},1,\frac{3}{2};\frac{1}{2}\rangle$ discussed above. Since $m_N = \frac{1}{2}$ for both states, we use two detectors with filters transmitting σ^- polarized light and one with a filter transmitting σ^+ polarized light. Further, in both cases we have $S_2 > S_1$ which implies that the first and the second emitter are connected to all three detectors. For the state $|\frac{1}{2},1,\frac{1}{2};\frac{1}{2}\rangle$, we find $S_3 < S_2$. Therefore, we connect the third emitter only to two detectors with filters transmitting σ^- and σ^+ polarized light, respectively, e.g., detector D_2 and D_3, and we

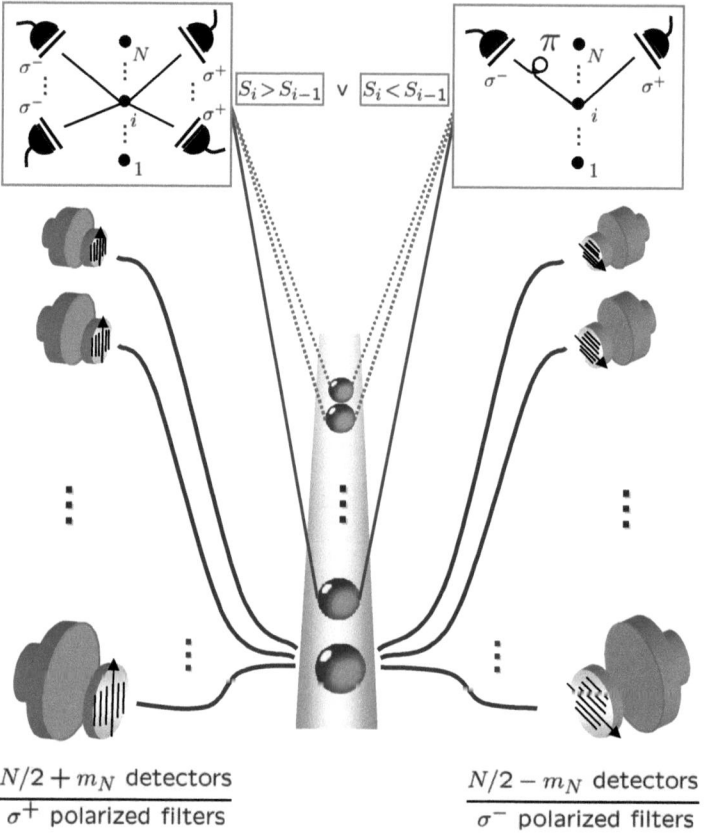

Figure 5.6: Experimental setup for mimicking the spin-spin coupling of N remote atoms via projective measurements. Depending on the coupling history of a particular N-qubit total angular momentum eigenstate, i.e., on the values of $S_1,..., S_N$, one has to modify the setup accordingly.

introduce an optical phase shift of π for the path leading from the third emitter to detector D_3. Summarizing we obtain the setup shown in figure 5.5 as postulated. For the state $|\frac{1}{2},1,\frac{3}{2};\frac{1}{2}\rangle$, we find $S_3 > S_2$. Here, we connect the third emitter to all three detectors. In this case, as mentioned above, the setup will generate the symmetric W-state [21].

5.3.4 Conclusions

Let us conclude this section. We considered a system of N remote non-interacting single-photon emitters with a Λ-level structure. By mimicking the coupling of angular momentum, we showed that it is possible to engineer any of the 2^N total angular momentum eigenstates in the ground-state qubits of this N-partite system [24].

Using linear optical tools only, our method employs the detection of all N photons scattered from the N emitters at N distinct polarization sensitive detectors. Thereby, it offers access to any of the 2^N states of the coupled basis of an N qubit compound system. Using projective measurements we thereby form highly and weakly entangled quantum states even though no interaction between the qubits is present [24]. For the technical feasibility of our scheme we refer to section 5.2.7.

5.4 Generation of symmetric entangled states by tuning of local operations

In sections 5.2 and 5.3, we showed how to generate symmetric and non-symmetric total angular momentum eigenstates [21, 24]. In the last part of chapter 5, we want to generalize our scheme of measurement based projection of entangled multi-partite quantum states towards the generation of states belonging to other *classes* of entanglement. We will demonstrate how the basic scheme introduced in section 5.1 can be used to determine entanglement classes themselves by simple tuning of easily accessible experimental parameters of the setup [22].

5.4.1 Introduction: tripartite entanglement classes of W, GHZ and separable states

For the case of two-qubit systems, entanglement is well understood and can be precisely quantified [101]. Three-qubit systems, apart from the trivial disentangled case (class of separable states), possess two inequivalent genuine tripartite entanglement classes [102, 124]. Entanglement classes of four qubits have been recently considered [103, 104], and efforts have been invested to proceed towards higher numbers of qubits [125], including an inductive method [126]. However, so far no comprehensive and scalable classification has been developed.

In the following, we will speak mainly about the tripartite class; here, as in section 5.3, the states $|+\rangle$ and $|-\rangle$ will define a qubit. The three tripartite classes are:

- the genuine entangled class of GHZ-type quantum states [127], which is represented by the maximally entangled state

$$|\text{GHZ}_N\rangle \equiv \frac{1}{\sqrt{2}}\left(e^{-i\phi}|+,\ldots,+\rangle_N + e^{i\phi}|-,\ldots,-\rangle_N\right), \quad (5.4.1)$$

with arbitrary relative phase ϕ;

- the class of separable (product) states of the form

$$|\text{S}_N\rangle \equiv |1_\phi,\ldots,1_\phi\rangle_N, \quad (5.4.2)$$

with $|1_\phi\rangle \equiv \left(|+\rangle + e^{i\phi}|-\rangle\right)/\sqrt{2}$;

- the class of multi-qubit states of the W type [102], represented by the weakly entangled state

$$|W_N\rangle = \frac{1}{\sqrt{N}}\Big(|1_\phi, 0_\phi, ..., 0_\phi\rangle_N + ... + |0_\phi, ..., 0_\phi, 1_\phi\rangle_N\Big), \qquad (5.4.3)$$

with $|0_\phi\rangle \equiv \big(|+\rangle - e^{i\phi}|-\rangle\big)/\sqrt{2}$.

Entanglement classes for the three-qubit case are well established [102, 124]: any three-qubit quantum state totally symmetric with respect to permutation of its qubits is separable or belongs either to the GHZ or the W family, though elements of the latter two classes do not always preserve the forms exhibited in Eqs. (5.4.1) and (5.4.3).

There are several physical systems where the generation of GHZ and W states with three or more qubits have been experimentally achieved: in trapped ions [8, 9], in Rydberg atoms crossing microwave cavities [128], and in photonic systems [129, 130]. Furthermore, other multi-qubit entangled states have been realized in different physical setups [69, 106, 107, 109, 110, 131, 132] with different purposes and potential applications in quantum information tasks. Though several paradigmatic entangled states have been produced in the lab, there is no study for the moment that associates operationally given experimental configurations with multi-partite entanglement classes of matter qubits[4].

Again, we will consider the measurement based scheme of projecting long-lived multi-qubit emitter states as introduced in section 5.1 and applied in sections 5.2 and 5.3. Here, we will show that it is possible to associate well-defined sets of locally tuned polarizer orientations with multi-qubit entanglement classes, allowing their monitoring in an operational manner. We will also argue that multi-path quantum interference, associated with qubit permutation symmetry, plays a key role in explaining the underlying physics.

5.4.2 Description of the physical system

We consider the basic setup introduced in section 5.1 for the case of N single-photon emitters aligned in a regular chain with equal next-neighbor distances. Each emitter defines a three-level Λ-system, where $|e\rangle$ denotes the excited state, while the two long-lived sublevels, $|+\rangle$ and $|-\rangle$, define a qubit. For the sake of simplicity, we assume again that the two transitions between the excited state and the two lower sublevels are equally probable, and that the accordingly emitted photons are circularly polarized, $\boldsymbol{\sigma}^-$ and $\boldsymbol{\sigma}^+$, respectively.

Figure 5.7 recalls the N-emitter case discussed throughout this section. All N emitters are initially excited and we will study the cases in which all spontaneously emitted photons are detected by N detectors D_j located in the far-field region, each of them being equipped

---

[4]Here, we want to point out that recent studies demonstrated an operational characterization of entanglement amongst photonic qubit states in an experiment [130].

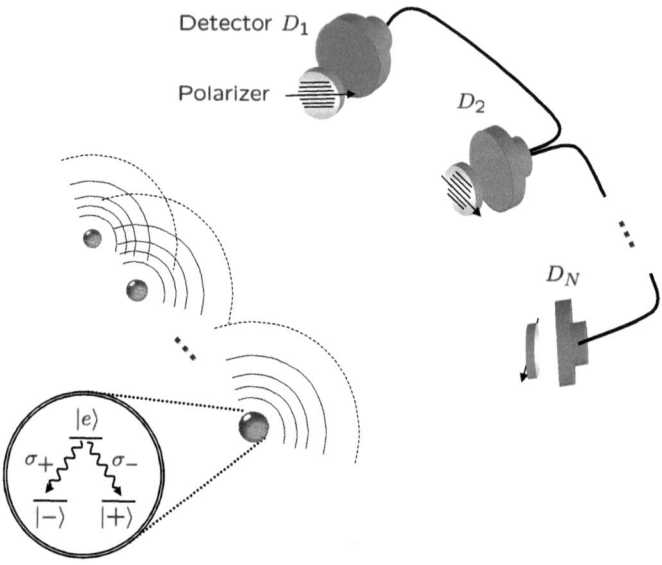

Figure 5.7: Proposed experimental arrangement. N excited emitters are aligned in a row, each of them defining a three-level Λ-system. In a successful measurement all N emitted photons are registered at N distinct detectors D_j placed in the far-field region of the emitters, each being equipped with a polarization filter in front transmitting only $\boldsymbol{\eta}_j$ polarized light (with $j = 1, ..., N$).

with a polarization filter in front transmitting only $\boldsymbol{\eta}_j$ polarized light (with $j = 1, ..., N$). We recall that the far-field detection ensures the erasure of which-way information of the arriving photons.

In analogy to section 5.2, we arrange the detectors at positions $\mathbf{r}_j$ ($j = 1, ..., N$) such that all optical pathways between emitters and detectors differ by multiples of 2π only. However, in contrast to sections 5.2 and 5.3, we now consider polarization filters which allow for projecting arbitrary elliptical polarizations in the x-y-plane of the circularly polarized light, i.e., where $\boldsymbol{\eta}_j = \alpha\,\boldsymbol{\sigma}^- + \beta\,\boldsymbol{\sigma}^+$ with $|\alpha|^2 + |\beta|^2 = 1$. As a consequence, a photon detection event projects the corresponding emitter onto a linear combination of the long-lived states $|-\rangle$ and $|+\rangle$ of the form $\alpha\,|+\rangle + \beta\,|-\rangle$. Therefore, the jth detector D_j implements the operator action

$$\hat{D}_N(\boldsymbol{\eta}_j) = \alpha_j \sum_{n=1}^{N} |+\rangle_n \langle e| + \beta_j \sum_{n=1}^{N} |-\rangle_n \langle e|, \qquad (5.4.4)$$

as derived from Eq. (5.1.4) (up to an insignificant normalization factor). Again $|\pm\rangle_n \langle e|$ denotes the atomic projection operator from the state $|e\rangle$ to the state $|\pm\rangle$ for the nth emitter. We note that we leave out $\mathbf{r}_j$ in the argument of the operator since we assume $\delta_j = n\, 2\pi$ for any j (n being an integer).

Starting with the N emitters in their excited state $\prod_{n=1}^{N} |e\rangle_n$, the detection of the N emitted photons at N detectors with polarizer configurations $\boldsymbol{\eta}_1, ..., \boldsymbol{\eta}_N$ projects the emitter system onto the final state $\prod_{j=1}^{N} \hat{D}_N(\boldsymbol{\eta}_{N-j+1}) \prod_{n=1}^{N} |e\rangle_n$ (c.f. Eq. (5.1.11)). Because of the symmetry properties of these operators, the final state as well as all intermediate states are totally symmetric with respect to permutations of the emitters. If entanglement is produced at the end of the detection process, we are thus ensured that only genuine multi-partite entangled states belonging to the accessible symmetric entanglement classes are generated.

5.4.3 Generation of 3-qubit W, GHZ and separable states

The intermediate states produced in the course of the successive photon detection events can be ordered in a pyramidal structure, displaying the various possible quantum paths towards the generation of the desired final state. Let us first illustrate the associated physics for the case of $N = 3$ qubits as shown in figure 5.8. At the end of the three photon detection process, the system is found in a coherent superposition of the eight possible product states of our three-qubit system in the $|\pm\rangle$ basis. All probability amplitudes related to the different quantum paths add up and yield interference terms according to the tagged prefactors of figure 5.8. The GHZ-type basis states, $|+,+,+\rangle$ and $|-,-,-\rangle$, located at the lower left and right corners of the pyramid, are the only final components (red circles) not subject to any quantum interferences as they can be obtained via a single quantum path (red arrow) only.

In order to generate the state of Eq. (5.4.1) with $N = 3$, $|\text{GHZ}_3\rangle$, it is thus required that *all* interference pathways add up simultaneously in a destructive manner, i.e., the interference terms displayed with blue and green circles in figure 5.8 must vanish. This condition is fulfilled if, e.g., linear polarizers $\boldsymbol{\eta}_j = \frac{1}{\sqrt{2}}(e^{-i\theta_j}\boldsymbol{\sigma}^+ + e^{i\theta_j}\boldsymbol{\sigma}^-)$ are used, i.e., setting $\alpha_j = \frac{1}{\sqrt{2}} e^{i\theta_j}$ and $\beta_j = \frac{1}{\sqrt{2}} e^{-i\theta_j}$, so that

$$\alpha_1\alpha_2\beta_3 + \alpha_1\alpha_3\beta_2 + \alpha_2\alpha_3\beta_1 = e^{i(\theta_1+\theta_2-\theta_3)} + e^{i(\theta_1+\theta_3-\theta_2)} + e^{i(\theta_2+\theta_3-\theta_1)} = 0,$$
$$\alpha_1\beta_2\beta_3 + \alpha_2\beta_1\beta_3 + \alpha_3\beta_1\beta_2 = e^{i(\theta_1-\theta_2-\theta_3)} + e^{i(\theta_2-\theta_1-\theta_3)} + e^{i(\theta_3-\theta_1-\theta_2)} = 0, \quad (5.4.5)$$

while setting $e^{i(\theta_1+\theta_2+\theta_3)}$ equal to $e^{i\phi}$. Such values for θ_j are given by the three third roots of $e^{i\phi}$, giving three complex numbers uniformly distributed on a circle of unit radius. The $|\text{GHZ}_3\rangle$ states in the computational $|\pm\rangle$ qubit basis are thus obtained when the three linear polarizers in front of the three detectors are adjusted such that

$$\theta_1 = \frac{\phi}{6}, \quad \theta_2 = \frac{\phi}{6} + \frac{\pi}{3}, \quad \theta_3 = \frac{\phi}{6} + \frac{2\pi}{3}. \quad (5.4.6)$$

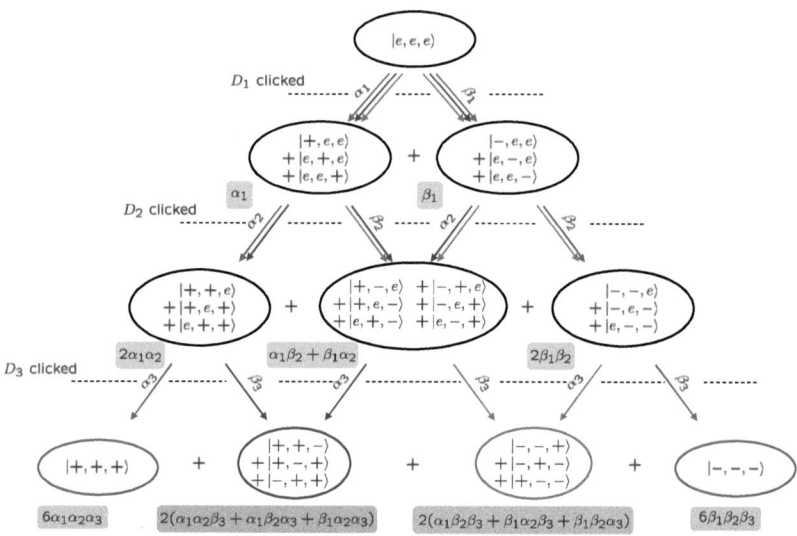

Figure 5.8: Pyramid of entanglement paths for the case of $N = 3$ emitters being initially in the excited state $|e,e,e\rangle$. The figure illustrates the intermediate states, horizontally displayed, during three successive photon detection events realized by three detectors D_1, D_2, and D_3, linearly polarized along η_1, η_2 and η_3 ($\eta_j \equiv \alpha_j \sigma^+ + \beta_j \sigma^-$), respectively. After each detection, the unnormalized global state of the system is the sum of all states in the circles weighed by the tagged prefactors. Left-down arrows denote σ^+ transitions, and right-down arrows represent σ^- ones. The final state components are shown inside colored frames and the colored arrows represent the quantum paths leading to these different components. Only the red circle states are obtained via a single quantum path, while the blue and green ones are the result of three different interfering quantum paths.

The same result is obtained for any permutation of the indices $j = \{1, 2, 3\}$ of θ_j. We remark that an entangled three-qubit state belonging to the GHZ class will be generated for any value of the relative phase ϕ, as can be see from Eq. (5.4.1).

In contrast, we are left with the three-qubit separable state of Eq. (5.4.2) with $N = 3$, $|S_3\rangle$, if all interference terms are chosen to add up simultaneously in a constructive manner, so that the tagged prefactors of the final state of figure 5.8 read successively 6, $6\,e^{i\phi}$, $6\,e^{i2\phi}$, and $6\,e^{i3\phi}$ (where we omitted an overall phase of $e^{i\frac{3}{2}\phi}$). Therefore all polarizers have to be oriented identically along the angle $\phi/2$ so that

$$\theta_1 = \theta_2 = \theta_3 = \frac{\phi}{2}. \qquad (5.4.7)$$

Note that the separable state $|S_3\rangle$ happens to fall in the rotated qubit basis, $|0_\phi\rangle$ and $|1_\phi\rangle$, as defined in Eq. (5.4.2), for any value of the phase ϕ.

Similarly, the state of Eq. (5.4.3) with $N = 3$, $|W_3\rangle$, is obtained when two polarizers are oriented identically along $\phi/2$ while the third one is chosen orthogonal to the first two,

$$\theta_1 = \theta_2 = \frac{\phi}{2}, \quad \theta_3 = \frac{\phi}{2} \pm \frac{\pi}{2}. \tag{5.4.8}$$

Again, the same result is obtained for any permutation of the indices $j = \{1, 2, 3\}$ of θ_j. Implementing this choice, a three-qubit state belonging to the W class is created for any value of the angle ϕ.

5.4.4 Generation of N-qubit W, GHZ and separable states

Let us generalize the scheme of the foregoing section for the N-qubit case: at the end of an N photon detection process, the system is found in a coherent superposition of all 2^N possible product states of the N-qubit system in the $|\pm\rangle$ basis. Again, all probability amplitudes related to the different quantum paths add up and yield interference terms. Here, the final state $|\Psi_N\rangle$ is found to be a linear combination of all symmetric Dicke states $|\frac{N}{2}, m_N\rangle$ [21]

$$|\Psi_N\rangle = \sum_{m_N=-\frac{N}{2}}^{\frac{N}{2}} c_m |\frac{N}{2}, m_N\rangle, \tag{5.4.9}$$

where $m := \frac{N}{2} + m_N$ defines the number of qubits projected onto the ground state $|+\rangle$, and

$$c_m = \binom{N}{m}^{1/2} \sum_{1 \leq j_1 \neq \ldots \neq j_N \leq N} \beta_{j_1} \ldots \beta_{j_m} \alpha_{j_{m+1}} \ldots \alpha_{j_N}. \tag{5.4.10}$$

Thereby, we can generate *any* symmetric state with respect to permutations of the emitters using suitable elliptical polarizer orientations. This is a direct application of Vieta's formulas[5]: an arbitrary symmetric state can be expanded in the symmetric Dicke state basis as $\sum_{m=0}^{N} d_m |\frac{N}{2}, m - \frac{N}{2}\rangle$ and is generated in our setup using M polarizers transmitting along elliptical polarization vectors $\boldsymbol{\eta}_j$ where we identify α_j/β_j with the M roots, i.e., the degree, of the polynomial $P(z) = \sum_{m=0}^{N}(-1)^{M-m}\sqrt{\binom{N}{m}/\binom{N}{M}}d_m z^m$. The remaining $N - M$ polarizers are oriented to transmit $\boldsymbol{\sigma}^+$ polarized light.

All final state components $|\frac{N}{2}, m_N\rangle$ $(-\frac{N}{2} < m_N < \frac{N}{2})$ except the GHZ-like states $|\frac{N}{2}, +\frac{N}{2}\rangle \equiv |+, \ldots, +\rangle_N$ and $|\frac{N}{2}, -\frac{N}{2}\rangle \equiv |-, \ldots, -\rangle$ are the result of many different quantum paths interfering with each other (compare with the sum term in Eq. (5.4.10)). This peculiar

---

[5]Vieta's formulas are sum-of-roots and product-of-roots identities for polynomials. See, e.g., E. W. Weisstein, *Vieta's Formulas*, MathWorld - A Wolfram Web Resource, http://mathworld.wolfram.com/VietasFormulas.html.

property gives these components a particular status: if *all* quantum path interferences are made simultaneously destructive or constructive with a judicious choice of the polarizer orientations the maximally-entangled state $|\text{GHZ}_N\rangle$ or the separable state $|S_N\rangle$ is generated at the end, respectively.

Considering, as before, linear polarizers $\boldsymbol{\eta}_j = \frac{1}{\sqrt{2}}(e^{-i\theta_j}\boldsymbol{\sigma}^+ + e^{i\theta_j}\boldsymbol{\sigma}^-)$, the maximally entangled state $|\text{GHZ}_N\rangle$ of Eq. (5.4.1) is generated if the polarizers are oriented along angles

$$\theta_j = \left[\frac{\pi}{2N}\right] + \frac{\phi}{2N} + (j-1)\frac{\pi}{N}, \quad j = 1, ..., N, \quad (5.4.11)$$

or any other configuration resulting from permuting the indices j of θ_j. Here, the term inside the square brackets is present only for the case of an even number of emitters. The state $|\text{GHZ}_N\rangle$ appears naturally when using all distinct polarizer orientations uniformly distributed over 2π.

In contrast, we are left with the separable state $|S_N\rangle$ of Eq. (5.4.2) with linear polarizers all identically oriented along the angle $\phi/2$ so that

$$\theta_1 = ... = \theta_N = \frac{\phi}{2}. \quad (5.4.12)$$

The state $|W_N\rangle$ is also generated with linear polarizers, all except one oriented identically along $\phi/2$ with the last orthogonal to the $N-1$ first,

$$\theta_1 = ... = \theta_{N-1} = \frac{\phi}{2}, \quad \theta_N = \frac{\phi}{2} \pm \frac{\pi}{2}, \quad (5.4.13)$$

or any configuration attainable by permutation of the indices.

5.4.5 Operational determination of tripartite entanglement classes

In section 5.4.3, analyzing the tripartite scenario of our setup, it turned out that the rotation of a single polarizer allows us to switch from one entanglement class to another one. These polarizer manipulations can be referred to as being local with respect to the (polarizer) detector positions, though they are not local with respect to the qubit positions: it is well known that one cannot convert states from different families into each other using stochastic local operations and classical communication (SLOCC)[6] only. However, remarkably, as illustrated in figure 5.9 for $N = 3$, the above described local polarization rotations enable the transitions

$$\text{S class} \leftrightarrow \text{W class} \leftrightarrow \text{GHZ class}, \quad (5.4.14)$$

---

[6]Only approximated conversions can be realized; see, e.g., [133].

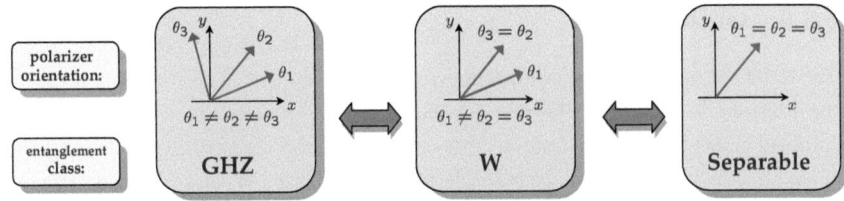

Figure 5.9: Operational association of polarization configurations with tripartite entanglement classes. If the three polarizer angles are all different, the final state belongs to the GHZ class; if two polarizer angles are different, the final state belongs to the W class; if all polarizers are identically oriented, the final state belongs to the S class.

where S stems from *separable*.

For three emitters, those particular results, associating the states $|S_3\rangle$, $|W_3\rangle$, and $|GHZ_3\rangle$, with certain choices of polarization angles, suggest a wider physical picture. In fact, we can associate in an operational manner specific polarizer configurations of our proposed physical setup with the three paradigmatic entanglement classes appearing in the three-qubit case: the number of distinct polarizer orientations identifies univocally the entanglement class to be S, W, or GHZ, as shown in figure 5.9.

According to Dür *et al.* [102], the GHZ class is formed by states characterized by a non-vanishing 3-tangle[7], the W class by states with a zero 3-tangle and non-zero single qubit von Neumann entropies, while the separable state class is characterized by zero values of these entanglement measures. In our proposed scheme, the 3-tangle τ of the final state $|\Psi_3\rangle$ reads [22]

$$\tau = \frac{4}{27} |\alpha_1\beta_2 - \alpha_2\beta_1|^2 |\alpha_1\beta_3 - \alpha_3\beta_1|^2 |\alpha_2\beta_3 - \alpha_3\beta_2|^2 \qquad (5.4.15)$$

and vanishes only when two polarizers are equally oriented. In this case, the local entropies vanish only when the third polarizer coincides with the two first [22]. When switching from a configuration with three distinct polarizer orientations to a configuration with three identical ones, via the intermediate case where two of them are equal, we transit thus successively from the GHZ class, to the W class, and end in the S class, as can be seen in figure 5.9. This shows the potential of the proposed setup for associating operationally a physical setting with three-qubit entanglement classes of states inequivalent under SLOCC and totally symmetric with respect to permutations of the emitters.

---

[7]The 3-tangle is a measure of genuine tripartite entanglement that is invariant under permutations. For a formal definition see [134].

5.4.6 Outlook

The results of section 5.4.5 encourage a possible generalization of the three-qubit case to the arbitrary N-qubit case. However, for the moment, we cannot say more about the operational entanglement classification of the general N-qubit case, with the only exception of the already described extremal entangled states: separable, W, and GHZ [22]. The reason is merely that, as we expressed before, recent studies have only made considerations up to the four-qubit case [104].

Chapter 6

Evidence and experimental proof of non-classicality

In the last part of this thesis we aim to analyze the underlying physics of the system introduced in chapter 2 in more detail and investigated throughout this work. In chapter 4, we showed how the correlations among the photons emitted by single photon sources can be exploited for an image processing technique which allows to obtain sub-classical resolution. In this context, we encountered correlations in the fluorescence light which are commonly attributed to the characteristics of entanglement. In particular, in section 3.4, we demonstrated that the two-photon correlation signal emitted by two initially excited two-level atoms has the potential to violate Bell-type inequalities and thus to prove the entangled nature of the observed correlations.

The present chapter will focus on the investigation of more involved inequalities which allow in particular for an experimentally feasible proof of the entangled nature of the fluorescence photons. Our considerations will be based on a slightly modified system of multi-level emitters similar to the one introduced in chapter 5. Hereby, we will investigate and exploit both degree of freedoms, i.e., the polarization degree of freedom and the position degree of freedom.

6.1 Historical introduction to Bell Inequalities

Bell-type inequalities like the ones presented in section 3.4 allow to test whether a two-partite signal measured independently at two different positions reveals correlations impossible to explain by means of a local deterministic theory. These inequalities can be traced back to Bell, who demonstrated in 1964 that quantum mechanics can give rise to results different from those which can be obtained from a classical local and deterministic theory [66].

For the derivation of his famous inequalities, Bell started from a classical theory with only three underlying assumptions, namely, the assumptions of a *local, deterministic* theory allowing for *hidden variables* (LHV). Obviously the latter additional degrees of freedom were inspired by the famous article of Einstein, Podolsky and Rosen written in 1935 [1]. Bell's initial inequalities stimulated further investigations which led to a number of inequalities that improved or extended the idea of testing the fundamentals of quantum mechanics. In particular, Clauser *et al.* formulated improved inequalities (so-called CHSH- or CH-type inequalities) which allowed to apply them in the context of a real experiment: in this approach, two correlated photons are emitted by an atomic cascade de-excitation process utilizing the polarization degree of freedom to demonstrate their entanglement which allowed to lower the experimental requirements [67, 68]. Thereafter, several groups demonstrated in a line of experiments that only the theory of quantum mechanics is capable of describing the corresponding non-local correlations among the polarization degrees of freedom of the photons [4, 135–137].

Today, Bell-type inequalities still provide an important tool to pinpoint the presence of entanglement in a given quantum system. Much effort has been spent on extending Bell- and CHSH-type inequalities so that it applies for example also for multi-partite quantum systems of arbitrary dimensions [138–140]. Here, we will use the CHSH-type Bell-inequalities to provide an experimentally feasible proof of the entangled nature of the two-photon correlations analyzed in chapter 4 and extend these investigations to a system of N single photon sources.

6.2 Investigating polarization correlations using CHSH inequalities

The experimental insufficiencies of Bell-type inequalities encountered in section 3.4 are well known since Bell's seminal paper in 1964 [66]. In 1969, Clauser *et al.* were able to improve the idea of testing quantum mechanical correlations by providing an experimentally feasible implementation [68]. In their paper, the authors considered pairs of photons emitted in an atomic cascade such that the polarization state of each photon is correlated with the polarization state of the other due to the conservation of angular momentum of the complete system. In the following we will demonstrate that a CHSH-type inequality can be formulated also for our particular system of two single photon emitters, considering polarization degrees of freedom and being independent of the overall success probability, when introducing multi-level atoms (compare with section 3.4).

6.2.1 Description of the physical system

We consider the setup shown in figure 6.1: two three-level systems with a V-level structure, i.e., three-level systems with two excited states $|e, -1\rangle$ and $|e, +1\rangle$ which both decay to a common ground state $|g, 0\rangle$ accompanied by the emission of a σ^+ or σ^- polarized photon, respectively (consider, e.g., Zeeman sub-levels), are located at $\mathbf{R}_1$ and $\mathbf{R}_2$. Initially, one atom is excited along the transition $|g, 0\rangle \to |e, +1\rangle$ and another one along $|g, 0\rangle \to |e, -1\rangle$. Hereby, it is precisely known which atom is initially in the $|e, +1\rangle$ state and which in the $|e, -1\rangle$ state. For the sake of simplicity we assume again that both transitions are equally probable.

As a consequence, the two tree-level emitters located at $\mathbf{R}_1$ and $\mathbf{R}_2$ will scatter exactly two photons with orthogonal polarizations σ^+ and σ^-, respectively. Again, we locate two detectors equipped with polarizers oriented along $\boldsymbol{\eta}_1$ and $\boldsymbol{\eta}_2$ at $\mathbf{r}_1$ and $\mathbf{r}_2$ in the far-field region of the emitters. In analogy to section 5.2.8, we define the wave vector $\mathbf{k}_j = k\mathbf{e}_j$ pointing into the direction of the detector at $\mathbf{r}_j$, where $\mathbf{e}_j := \frac{\mathbf{r}_j}{|\mathbf{r}_j|}$, and denote the corresponding spatial mode by $|k_j\rangle$. Via post-selection we assure that each of the two modes $|k_1\rangle$ and $|k_2\rangle$ is populated by exactly one photon. However, due to the measurement in the far-field region, no which-way information is obtained by the measurement of the photons and thus the quantum state of the two photons with orthogonal polarizations just before each detector registers a single-photon event can be written as

$$|k_1, k_2\rangle = \frac{1}{\sqrt{2}}(|\boldsymbol{\sigma}^+, \boldsymbol{\sigma}^-\rangle + |\boldsymbol{\sigma}^-, \boldsymbol{\sigma}^+\rangle). \tag{6.2.1}$$

This correlated quantum state is often associated with Bohm who used it in a first Gedankenexperiment to describe the phenomenon of non-classical correlations, as pointed

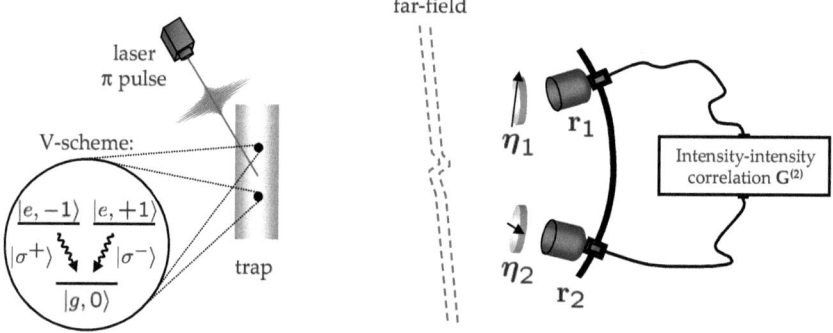

Figure 6.1: Setup used for testing CHSH-type inequalities: in order to introduce polarization degrees of freedom, we consider two atoms with a V-level structure, one being excited along the transition $|g, 0\rangle \to |e, +1\rangle$ and the other one along $|g, 0\rangle \to |e, -1\rangle$. The single-photon detectors placed at $\mathbf{r}_1$ and $\mathbf{r}_2$ in the far-field region of the emitters are equipped with polarization filters transmitting $\boldsymbol{\eta}_1$ and $\boldsymbol{\eta}_2$ polarized light, respectively.

out in the Einstein-Podolsky-Rosen article [1], in terms of polarization degrees of freedom [141, 142].

In analogy to the general detection operator of Eq. (5.1.4) which applies for photons emitted by N atoms with a Λ-level structure, we derive the following detection operator for a system of two emitters with a V-level structure: for the jth detector located at position $\mathbf{r}_j$ and being equipped with a polarization filter transmitting $\boldsymbol{\eta}_j$ polarized light we obtain in case of an initial state $|e, -1; e, +1\rangle$ (c.f. Eq. (2.2.13))

$$\hat{D}_2(\delta_j, \boldsymbol{\eta}_j) = \frac{\mathcal{E}_j}{\sqrt{2}} \left[(\boldsymbol{\eta}_j \cdot \boldsymbol{\sigma}^+)(|g,0\rangle\langle e,-1|) + e^{i\delta_j}(\boldsymbol{\eta}_j \cdot \boldsymbol{\sigma}^-)(|g,0\rangle\langle e,+1|) \right]. \qquad (6.2.2)$$

Please note that we omitted an overall phase factor of $e^{i\delta_j}$.

With the detection operator at hand, we can determine the joint detection probability for the system depicted in figure 6.1 by calculating the intensity correlation function of second order

$$G^{(2)}(\delta_1, \delta_2; \boldsymbol{\eta}_1, \boldsymbol{\eta}_2) = \left| \hat{D}_2(\delta_2, \boldsymbol{\eta}_2) \hat{D}_2(\delta_1, \boldsymbol{\eta}_1) |e, -1; e, +1\rangle \right|^2. \qquad (6.2.3)$$

In order to simplify the notation, we define $\boldsymbol{\eta}_j = \sin\vartheta_j \boldsymbol{\sigma}^+ + \cos\vartheta_j \boldsymbol{\sigma}^-$ so that $(\boldsymbol{\eta}_j \cdot \boldsymbol{\sigma}^+) = \sin\vartheta_j$ and $(\boldsymbol{\eta}_j \cdot \boldsymbol{\sigma}^-) = \cos\vartheta_j$. Moreover, the normalization of the detection operator becomes $\mathcal{E}_j^2 := \mathcal{E}_0^2/(|(\boldsymbol{\eta}_j \cdot \boldsymbol{\sigma}^-)|^2 + |(\boldsymbol{\eta}_j \cdot \boldsymbol{\sigma}^+)|^2) = \mathcal{E}_0^2$ (c.f. Eq. (5.1.5)).

Using these notations Eq. (6.2.2) reads

$$\hat{D}_2(\delta_j, \boldsymbol{\eta}_j) = \frac{\mathcal{E}_0}{\sqrt{2}} \left(\sin \vartheta_j |g, 0\rangle\langle e, -1| + e^{i\delta_j} \cos \vartheta_j |g, 0\rangle\langle e, +1| \right), \quad (6.2.4)$$

and from Eqs. (6.2.3) and (6.2.4), we obtain for the intensity correlation function of second order

$$G^{(2)}(\delta_1, \delta_2; \vartheta_1, \vartheta_2) = \frac{\mathcal{E}_0^4}{4} \left| e^{i\delta_2} \sin \vartheta_1 \cos \vartheta_2 + e^{i\delta_1} \sin \vartheta_2 \cos \vartheta_1 \right|^2. \quad (6.2.5)$$

As there is a unique correspondence between $\boldsymbol{\eta}_j$ and ϑ_j, we can write Eq. (6.2.5) as a function of ϑ_1 and ϑ_2.

6.2.2 Derivation of CHSH inequalities for polarization correlations

Let us start to analyze the polarization degrees of freedom by fixing the positions of the two detectors in such a way that all optical pathways leading from any of the two atoms to each of the two detectors are identical up to phase differences of odd multiples of π[1]. Hence, the polarization axes $\boldsymbol{\eta}_1$ and $\boldsymbol{\eta}_2$ are the only free parameters for the variation of the intensity correlation function of second order derived in Eq. (6.2.5).

For the particular choice of the two detector positions such that $\delta_2 = \alpha 2\pi$ and $\delta_1 = (2\alpha + 1)\pi$ with α being an integer, the intensity correlation function of second order becomes

$$G^{(2)}(\vartheta_1, \vartheta_2) = \frac{\mathcal{E}_0^4}{4} \sin^2 (\vartheta_2 - \vartheta_1). \quad (6.2.6)$$

With Eq. (6.2.6) at hand, we can determine the joint detection probability for the system shown in figure 6.1 as

$$P(\vartheta_1, \vartheta_2) = \frac{C_0^2}{\mathcal{E}_0^4} G^{(2)}(\vartheta_1, \vartheta_2) = \frac{C_0^2}{4} \sin^2 (\vartheta_2 - \vartheta_1). \quad (6.2.7)$$

In a next step we will use the joint detection probability of Eq. (6.2.7) to derive adequate CHSH-type inequalities which depend solely on the polarization degrees of freedom.

Again, we start our analyses by considering the inequalities (3.4.3). However, instead of considering spatial correlations as in section 3.4, we want to investigate correlations among the photons' polarization degree of freedom. Hence, we focus on detection probabilities of the form $p(\vartheta_j, \lambda)$. In [68] Clauser et al. identified the parameters in the inequalities (3.4.3)

---

[1] We note that this choice is totally arbitrary. However, the particular advantage of a phase shift of π is to obtain a nice analytical expression as in Eq. (6.2.6).

similar to the following set of detection probabilities

$$p(\vartheta_1, \lambda) = x, \quad p(\vartheta_1', \lambda) = x', \quad p(\infty, \lambda) = X,$$
$$p(\vartheta_2, \lambda) = y, \quad p(\vartheta_2', \lambda) = y', \quad p(\infty, \lambda) = Y. \tag{6.2.8}$$

Here, $\vartheta_j = \infty$ denotes the particular case, where the polarization filter is removed (thereby allowing for all possible polarizations to be transmitted). The identification in the last column of Eq. (6.2.8) guarantees the constraint $X \geq x, x'$ ($Y \geq y, y'$), assumed in the derivation of (3.4.3), which is justified by the *no-enhancement* condition [59, 67, 68]: the detection probability of a measurement performed with polarization filter cannot exceeded the measurement without a polarization filter.

According to the requirements of a LHV theory (c.f. Eq. (3.4.1)) the joint probability of detecting both photons is written as

$$p_{12}(\vartheta_1, \vartheta_2, \lambda) = p(\vartheta_1, \lambda) p(\vartheta_2, \lambda). \tag{6.2.9}$$

Using Eqs. (6.2.8) and (6.2.9) in the inequalities (3.4.3), we obtain, after multiplying the inequalities by $g(\lambda)$ and integrating over λ,

$$-p_{12}(\infty, \infty) \leq \ p_{12}(\vartheta_1, \vartheta_2) - p_{12}(\vartheta_1, \vartheta_2') + p_{12}(\vartheta_1', \vartheta_2)$$
$$+ p_{12}(\vartheta_1', \vartheta_2') - p_{12}(\vartheta_1', \infty) - p_{12}(\infty, \vartheta_2) \leq 0. \tag{6.2.10}$$

Finally, dividing (6.2.10) by $p_{12}(\infty, \infty)$ we obtain the CHSH-type inequalities

$$-1 \leq \ [p_{12}(\vartheta_1, \vartheta_2) - p_{12}(\vartheta_1, \vartheta_2') + p_{12}(\vartheta_1', \vartheta_2) \tag{6.2.11}$$
$$+ p_{12}(\vartheta_1', \vartheta_2') - p_{12}(\vartheta_1', \infty) - p_{12}(\infty, \vartheta_2)]/p_{12}(\infty, \infty) \leq 0,$$

which we will apply to our setup in the next section.

6.2.3 Violating CHSH inequalities by polarization correlations

In difference to the Bell-type inequalities (3.4.5), the CHSH-type inequalities (6.2.11) depend on joint detection probabilities only. Moreover, they are normalized by the reference value $p_{12}(\infty, \infty)$, i.e., the joint detection probability where both polarizers are removed. With $G^{(2)}(\vartheta_1, \vartheta_2)$ of Eq. (6.2.6) at hand and by using the relation $p_{12}(\vartheta_1, \infty) = p_{12}(\vartheta_1, \vartheta_2) +$

$p_{12}(\vartheta_1, \vartheta_2 + \frac{\pi}{2})$, we find for the three different joint detection probabilities (c.f. Eq. (2.2.9))

$$p_{12}(\vartheta_1, \vartheta_2) \equiv P(\vartheta_1, \vartheta_2) = \mathcal{C}_0^2 \frac{1}{4} \sin^2(\vartheta_1 - \vartheta_2), \qquad (6.2.12)$$

$$p_{12}(\vartheta_1, \infty) \equiv P(\vartheta_1, \infty) = \mathcal{C}_0^2 \frac{1}{4}, \qquad (6.2.13)$$

$$p_{12}(\infty, \infty) \equiv P(\infty, \infty) = \mathcal{C}_0^2 \frac{1}{2}. \qquad (6.2.14)$$

Using these results together with the CHSH-type inequalities (6.2.11), we find

$$\begin{aligned} -1 \leq \; & \tfrac{1}{2}\sin^2(\vartheta_2 - \vartheta_1) - \tfrac{1}{2}\sin^2(\vartheta_2' - \vartheta_1) \\ & + \tfrac{1}{2}\sin^2(\vartheta_2 - \vartheta_1') + \tfrac{1}{2}\sin^2(\vartheta_2' - \vartheta_1') \; -1 \leq 0. \end{aligned} \qquad (6.2.15)$$

Finally, looking for the extrema of (6.2.15), we introduce again the sets of parameters of Eqs. (3.4.8) and (3.4.9). Substituting δ_j for ϑ_j, the two sets read

$$a.) = \begin{cases} \vartheta_2 - \vartheta_1 = \tfrac{3}{8}\pi, & \vartheta_2' - \vartheta_1 = \tfrac{1}{8}\pi, \\ \vartheta_2 - \vartheta_1' = \tfrac{3}{8}\pi, & \vartheta_2' - \vartheta_1' = \tfrac{3}{8}\pi, \end{cases} \qquad (6.2.16)$$

$$b.) = \begin{cases} \vartheta_2 - \vartheta_1 = \tfrac{1}{8}\pi, & \vartheta_2' - \vartheta_1 = \tfrac{3}{8}\pi, \\ \vartheta_2 - \vartheta_1' = \tfrac{1}{8}\pi, & \vartheta_2' - \vartheta_1' = \tfrac{1}{8}\pi. \end{cases} \qquad (6.2.17)$$

Using these sets in (6.2.15), we obtain

$$-1 \leq \frac{\pm\sqrt{2} - 1}{2} \leq 0, \qquad (6.2.18)$$

where the plus sign holds for the set of Eq. (6.2.16), violating the upper bound of the inequality, and the minus sign holds for the set of Eq. (6.2.17), violating the lower bound. In conclusion, for the system shown in figure 6.1, the CHSH-type inequalities (6.2.11) or (6.2.18) can be clearly violated by inserting the polarization dependent joint detection probabilities of Eqs. (6.2.12)-(6.2.14). The violation of CHSH-type inequalities is commonly accepted as a proof for the entangled nature of the correlation signal measured.

The CHSH-type inequalities for polarization correlations are advantageous in comparison with the Bell-type inequalities introduced before (c.f. section 3.4) for several reasons: first, as can be seen from (6.2.11), only joint detection probabilities, which are subject to the same overall success probability, have to be measured. Moreover, due to the normalization, realized by dividing the inequalities by a reference value, they become independent of the overall success probability. Thereby, it is possible to circumvent any experimental insufficiencies which would lower some of the probabilities of Eqs. (6.2.12)-

(6.2.14). As a result the violation of these type of inequalities can be measured within a realistic experimental environment as the remaining sources of errors, e.g., the localization of the atomic emitters, can be controlled to a great extend (see, e.g., [143]).

6.3 Investigating spatial correlations using CHSH inequalities

In this section we apply the CHSH-type inequalities to the setup shown in figure 6.1 where we now include all degrees of freedom, i.e., polarization *and* position variables. While in sections 6.2.2 and 6.2.3 we considered the state of Eq. (6.2.1) as being the photonic state registered by the detectors, we did not account for any spatial correlations of the photons.

6.3.1 Derivation of CHSH inequalities for spatial correlations

Following the derivation of the CHSH-type inequalities considered in section 6.2.2, we start out again with the mathematical inequalities (3.4.3). This time we identify the detection probabilities as follows

$$p(\delta_1,\vartheta_1,\lambda)=x,\ p(\delta_1',\vartheta_1,\lambda)=x',\ p(\delta_1,\infty,\lambda)=X,$$
$$p(\delta_2,\vartheta_2,\lambda)=y,\ p(\delta_2',\vartheta_2,\lambda)=y',\ p(\delta_2,\infty,\lambda)=Y, \tag{6.3.1}$$

where ∞ indicates that the polarization filter is removed for the particular measurement[2]. Please note that we included the polarization degree of freedom ϑ_j as well as the position degree of freedom δ_j in the argument of the probabilities.

Again, following the assumptions of a local deterministic theory with hidden variables λ, we can write the joint probability as

$$p_{12}(\delta_1,\delta_2;\vartheta_1,\vartheta_2,\lambda) = p(\delta_1,\vartheta_1,\lambda) \cdot p(\delta_2,\vartheta_2,\lambda). \tag{6.3.2}$$

Using Eqs. (6.3.1) and (6.3.2) in the inequalities (3.4.3), we obtain, after multiplying the whole relation with $g(\lambda)$ and integrating over λ,

$$\begin{aligned}-1 \leq\ & [p_{12}(\delta_1,\delta_2;\vartheta_1,\vartheta_2) - p_{12}(\delta_1,\delta_2';\vartheta_1,\vartheta_2) \\ & + p_{12}(\delta_1',\delta_2;\vartheta_1,\vartheta_2) + p_{12}(\delta_1',\delta_2';\vartheta_1,\vartheta_2) \\ & - p_{12}(\delta_1',\delta_2;\vartheta_1,\infty) - p_{12}(\delta_1,\delta_2;\infty,\vartheta_2)]/p_{12}(\delta_1,\delta_2;\infty,\infty) \leq 0.\end{aligned} \tag{6.3.3}$$

These are the CHSH-type inequalities suited for proving the quantum nature of the spatial correlations in the two-photon signal obtained with the setup depicted in figure 6.1.

---

[2]Again, the constraint $X \geq x, x'$ ($Y \geq y, y'$) is guaranteed by the *no-enhancement* condition [59,67,68]: the detection probability with a polarization filter cannot exceeded the measurement without a polarization filter.

6.3.2 Violating CHSH inequalities by spatial correlations

The introduction of the polarization degrees of freedom allowed us to derive the CHSH-type inequalities (6.3.3) which contain only joint detection probabilities (compare with Eq. (3.4.5)). However, in this section we want to focus on the spatial correlations only. Therefore, we adjust the polarizers such that $\vartheta_1 = \pi/4$ and $\vartheta_2 = -\pi/4$ so as to optimize the detection efficiency of the experimental setup. Implementing this constraint Eq. (6.2.6) becomes

$$G^{(2)}(\delta_1, \delta_2; \pi/4, -\pi/4) = \frac{\mathcal{E}_0^4}{8}\left(1 - \cos[2(\delta_1 - \delta_2)]\right). \quad (6.3.4)$$

However, as discussed, e.g., in section 4.2.4, experimental uncertainties tend to decrease the visibility of the correlation signal. Therefore, let us introduce artificially a visibility $\mathcal{V}$ of the second order correlation signal which allows to estimate the experimental requirements needed for a successful application of the CHSH-type inequalities. Eq. (6.3.4) is then substituted by the following expression

$$G^{(2)}(\delta_1, \delta_2; \pi/4, -\pi/4) = \frac{\mathcal{E}_0^4}{8}\left(1 - \mathcal{V}\cos[2(\delta_1 - \delta_2)]\right). \quad (6.3.5)$$

Next, having determined $G^{(2)}(\delta_1, \delta_2; \pi/4, -\pi/4)$, we can calculate the joint detection probabilities for the setup shown in figure 6.1. Employing again the relation $p_{12}(\delta_1, \delta_2; \vartheta_1, \infty) = p_{12}(\delta_1, \delta_2; \vartheta_1, \vartheta_2) + p_{12}(\delta_1, \delta_2; \vartheta_1, \vartheta_2 + \pi/2)$ and using Eq. (2.2.9), we identify

$$p_{12}(\delta_1, \delta_2; \pi/4, -\pi/4) \equiv P(\delta_1, \delta_2; \pi/4, -\pi/4) = \mathcal{C}_0^2 \frac{1}{8}(1 - \mathcal{V}\cos[2(\delta_1 - \delta_2)]), \quad (6.3.6)$$

$$p_{12}(\delta_1, \delta_2; \vartheta_1, \infty) \equiv P(\delta_1, \delta_2; \vartheta_1, \infty) = \mathcal{C}_0^2 \frac{1}{4}, \quad (6.3.7)$$

$$p_{12}(\delta_1, \delta_2; \infty, \infty) \equiv P(\delta_1, \delta_2; \infty, \infty) = \mathcal{C}_0^2 \frac{1}{2}. \quad (6.3.8)$$

Using these results together with the CHSH-type inequalities derived in (6.3.3), we find

$$-1 \leq \tfrac{1}{4}(1 - \mathcal{V}\cos[2(\delta_1 - \delta_2)]) - \tfrac{1}{4}(1 - \mathcal{V}\cos[2(\delta_1 - \delta_2')]) \\ + \tfrac{1}{4}(1 - \mathcal{V}\cos[2(\delta_1' - \delta_2)]) + \tfrac{1}{4}(1 - \mathcal{V}\cos[2(\delta_1' - \delta_2')]) - 1 \leq 0. \quad (6.3.9)$$

Finally, looking for the extrema of Eq. (6.3.9), we use again the two sets of parameters introduced in Eqs. (3.4.8) and (3.4.9). In combination with (6.3.9), we thus obtain the CHSH-type inequalities with respect to the spatial correlations of the emitted photons

$$-1 \leq \frac{\pm \mathcal{V}\sqrt{2} - 1}{2} \leq 0, \quad (6.3.10)$$

where the plus sign holds for the set of Eq. (3.4.8), violating the upper bound of the inequality if $\mathcal{V} > 1/\sqrt{2}$, and the minus sign holds for the set of Eq. (3.4.9), violating the lower bound of the inequality if $\mathcal{V} > 1/\sqrt{2}$.

In conclusion, the violation of the CHSH-type inequalities (6.3.10) has been demonstrated by use of the position dependent joint detection probabilities of Eqs. (6.3.6)-(6.3.8). For an experimental demonstration, it turns out to be crucial that a visibility $\mathcal{V} > 1/\sqrt{2} \approx 71\%$ can be achieved. In contrast to the experimental requirements needed to violate the Bell-type inequalities (3.4.10), the experimental requirements to violate the CHSH-type inequalities (6.3.10), i.e., achieving a visibility $\mathcal{V} > 1/\sqrt{2} \approx 71\%$, are realistic and feasible. Moreover, the overal success probability plays no longer a crucial role (see, e.g., [143]).

Let us emphasize that a violation of the CHSH-type inequalities (6.3.10) is considered commonly as a proof that the signals, i.e., the two photons measured at different positions, are entangled. In the next section, further implications resulting from this violation will be derived. There, we will look at the time dependencies of the joint detection events as derived in sections 2.1.5 and 2.2, which were excluded so far in our investigations[3].

---

[3]As pointed out in section 2.2 the time dependence can be neglected when considering coincident detection events only. Further, if the atoms are separated by an inter atomic distance $d \gg \lambda$, the time dependence of the respective intensity correlation functions factorizes (c.f. section 6.4 for further details).

6.4 Time dependent spatial correlations

So far we considered CHSH-type inequalities employing joint detection probabilities for the system shown in figure 6.1. Thereby, we assumed coincident detection of the photons. In contrast, here we turn our attention to an interpretation of the CHSH-type inequalities where we take the time dependences of the intensity correlation functions as derived in section 2.2 explicitly into account.

6.4.1 Time dependent intensity correlations of second order

In section 2.2 we derived an analytic expression for the time dependent intensity correlation function of Nth order (c.f. Eq. (2.2.8)). For $d \gg \lambda$ we found that this expression factorizes into an exponential function decreasing with time and the intensity correlation function of Nth order, determined under the assumptions of coincident detection events. In the following, we assume $d \gg \lambda$ for all derivations. In particular for $N = 2$ we found

$$G^{(2)}(\mathbf{r}_1, \mathbf{r}_2; t_1, t_2) = \prod_{j=1}^{2} e^{-2\gamma\left(t_j - \frac{|\mathbf{r}_j|}{c}\right)} G^{(2)}(\mathbf{r}_1, \mathbf{r}_2). \quad (6.4.1)$$

Hence, we see that the overall success probability of a particular joint detection event will dependent on both measurements, i.e., on the time for detecting the first and the time for detecting the second scattered photon, respectively. To simplify the situation we will make use of the common far-field approximation $|\mathbf{r}_1| \approx |\mathbf{r}_2|$ in the following.

6.4.2 Derivation and violation of CHSH inequalities for time dependent spatial correlations

Starting from Eq. (6.4.1) we briefly repeat our analyses of CHSH-type inequalities. Accounting explicitly for the time dependences, the joint detection probabilities of Eqs. (6.3.6)-(6.3.8) become

$$p_{12}(\delta_1, \delta_2; t_1, t_2; \pi/4, -\pi/4) = C_0^2 \frac{1}{8} e^{-2\gamma(t_1 + t_2 - 2\frac{|\mathbf{r}_1|}{c})} (1 - \mathcal{V} \cos[2(\delta_1 - \delta_2)]), \quad (6.4.2)$$

$$p_{12}(\delta_1, \delta_2; t_1, t_2; \vartheta_1, \infty) = C_0^2 \frac{1}{4} e^{-2\gamma(t_1 + t_2 - 2\frac{|\mathbf{r}_1|}{c})}, \quad (6.4.3)$$

$$p_{12}(\delta_1, \delta_2; t_1, t_2; \infty, \infty) = C_0^2 \frac{1}{2} e^{-2\gamma(t_1 + t_2 - 2\frac{|\mathbf{r}_1|}{c})}. \quad (6.4.4)$$

Next, we recall (6.3.3) which now reads

$$\begin{aligned}-1 \leq\ & [p_{12}(\delta_1, \delta_2; t_1, t_2; \vartheta_1, \vartheta_2) - p_{12}(\delta_1, \delta_2'; t_1, t_2; \vartheta_1, \vartheta_2) \\ & + p_{12}(\delta_1', \delta_2; t_1, t_2; \vartheta_1, \vartheta_2) + p_{12}(\delta_1', \delta_2'; t_1, t_2; \vartheta_1, \vartheta_2) \\ & - p_{12}(\delta_1', \delta_2; t_1, t_2; \vartheta_1, \infty) - p_{12}(\delta_1, \delta_2; \infty, \vartheta_2)]/p_{12}(\delta_1, \delta_2; t_1, t_2; \infty, \infty) \leq 0,\end{aligned} \qquad (6.4.5)$$

where we restrict the measurements only to those cases where the two photons are detected after uniquely preselected arrival times t_1 and t_2. We note that this requirement can be easily fulfilled in an experiment with a pulsed excitation scheme by post-selecting the appropriate results.

Again, by combining the inequalities (6.4.5) with the joint detection probabilities of Eqs. (6.4.2)-(6.4.4), we find

$$\begin{aligned}-1 \leq\ & \tfrac{1}{4}(1 - \mathcal{V}\cos[2(\delta_1 - \delta_2)]) - \tfrac{1}{4}(1 - \mathcal{V}\cos[2(\delta_1 - \delta_2')]) \\ & + \tfrac{1}{4}(1 - \mathcal{V}\cos[2(\delta_1' - \delta_2)]) + \tfrac{1}{4}(1 - \mathcal{V}\cos[2(\delta_1' - \delta_2')]) - 1 \leq 0,\end{aligned} \qquad (6.4.6)$$

which are essentially the same CHSH-type inequalities as were derived in (6.3.9). Thus, by using either of the two sets of parameters given in Eqs. (3.4.8) and (3.4.9) we find

$$-1 \leq \frac{\pm \mathcal{V}\sqrt{2} - 1}{2} \leq 0, \qquad (6.4.7)$$

where the plus sign holds for the set of Eq. (3.4.8), violating the upper bound of the inequality if $\mathcal{V} > 1/\sqrt{2}$, and the minus sign holds for the set of Eq. (3.4.9), violating the lower bound of the inequality if $\mathcal{V} > 1/\sqrt{2}$.

6.4.3 Interpretation

According to (6.4.5) - (6.4.7), we find that the CHSH-type inequalities are violated even when considering non-coincident detection events of the two photons contributing to the joint detection probabilities. It turned out that the time dependence cancels out when considering CHSH-type inequalities so that the violation remains for visibilities $\mathcal{V} > 71\%$, as demonstrated in (6.3.10) or (6.4.7), independent of the arrival times t_1 and t_2 of the photons.

For the considered system, this time independence gives rise to some very interesting Gedankenexperiments. In fact, the arbitrary photon arrival times t_1 and t_2 include the following scenario: let us assume that a first photon event is registered after a time t_1 at a detector at $\mathbf{r}_1$. Then, a second photon might be detected at $\mathbf{r}_2$ and time t_2 where we choose $t_2 > t_1 + \frac{|\mathbf{r}_2|}{c}$, i.e., via post-selection, we restrict the measurement only to those detection events where $t_2 > t_1 + \frac{|\mathbf{r}_2|}{c}$. In this scenario the two photons do not even exist in the same time interval (there is no overlap of their propagation times). Nevertheless, we do obtain a modulated joint detection probability as displayed in Eq. (6.4.2) and in this way the possibility to violate the CHSH-type inequality (c.f. (6.3.10)).

First, we explain why the modulated joint detection probability is preserved in case the two photons do not share the same time interval of existence. As shown in figure 6.1, initially the two atoms in our scheme are excited to a state $|e, -1\rangle$ and $|e, +1\rangle$ while both decay to the same ground state, i.e., to $|g, 0\rangle$, accompanied by the emission of a σ^+ or a σ^- polarized photon, respectively. After a time t_1 we measure a first scattered photon at a detector located in the far-field of the emitters at $\mathbf{r}_1$ and equipped with a polarizer assumed to be oriented along $\boldsymbol{\eta}_1 = 1/\sqrt{2}(\boldsymbol{\sigma}^+ + \boldsymbol{\sigma}^-)$ so that no which-way information can be retrieved from the measurement of the polarization of the photon. Since we know that $t_2 > t_1 + \frac{|\mathbf{r}_2|}{c}$, we might write down the (unnormalized) intermediate state of the atomic emitters which is obtained due to projection after the detection of a photon at $\mathbf{r}_1$ as (see Eq. (5.1.10))

$$\hat{D}_2(\delta_1, \boldsymbol{\eta}_1)|e, -1; e, +1\rangle \sim |g, 0; e, +1\rangle + e^{i\delta_1}|e, -1; g, 0\rangle. \quad (6.4.8)$$

We note that the intermediate state of Eq. (6.4.8) is projected only *after* the detection of the first scattered photon, i.e., after the first photon has been absorbed and *destroyed*. Now, if a second photon is scattered by the state of Eq. (6.4.8) and detected after a time $t_2 > t_1 + \frac{|\mathbf{r}_2|}{c}$ at a detector located in the far-field of the emitters at $\mathbf{r}_2$ and equipped with a polarizer oriented along $\boldsymbol{\eta}_2 = 1/\sqrt{2}(\boldsymbol{\sigma}^+ - \boldsymbol{\sigma}^-)$, then, again, from the measurement it is impossible to retrieve any which-way information. As a consequence, both terms of the state (6.4.8) contribute to the measurement. In particular, the second term is shifted by a phase $e^{i\delta_1}$ with respect to the first term, which gives rise to the modulated detection probability of the second scattered photon which thus depends on the position $\mathbf{r}_1$ of where the first photon was detected (c.f. Eq. (6.4.2)). Hereby, the state (6.4.8) can be seen as a *quantum memory*: despite the non-existence of the first photon due to its destructive detection at $\mathbf{r}_1$ after a time t_1, the information of this detection event is encoded and preserved within the optical phase δ_1 of the intermediate atomic state of Eq. (6.4.8).

The modulated joint detection probability, in case that $t_2 > t_1 + \frac{|\mathbf{r}_2|}{c}$, can be explained satisfactorily when using the language of quantum paths discussed, e.g., in section 3.2. Here, we define by $|0\rangle_1|0\rangle_2$ that none of the photons was scattered, by $|k\rangle_1|0\rangle_2$ that one photon was scattered and registered at a detector located at $\mathbf{r}_1$ ($\mathbf{r}_2$) and by $|k\rangle_1|k\rangle_2$ that both photons were registered at $\mathbf{r}_2$ ($\mathbf{r}_1$). For the considered case, the whole system can evolve along two different quantum paths, namely,

1. *quantum path:*

$$|e, -1; e, +1\rangle \otimes |0\rangle_1|0\rangle_2 \quad \to \quad |g, 0; e, +1\rangle \otimes |k\rangle_1|0\rangle_2 \quad \to \quad |g, 0; g, 0\rangle \otimes |k\rangle_1|k\rangle_2,$$

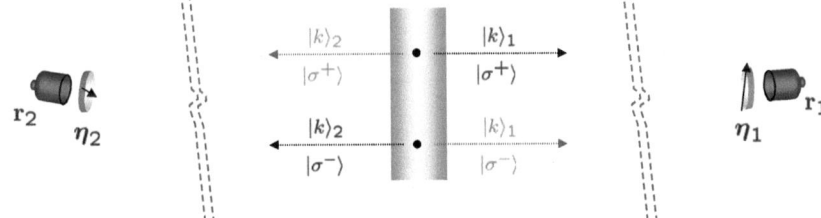

Figure 6.2: Setup capable of performing a time-loophole-free test of the CHSH-type inequalities: an LHV theory can restore a quantum mechanical correlation by assuming a communication between the detector at $\mathbf{r}_1$ and the one at $\mathbf{r}_2$. However, the transmission speed of such a communication would be limited by the speed of light c so that the transmission time between the two detectors, for the setup displayed, is given by $2\frac{|\mathbf{r}_1|}{c}$ at most (for $|\mathbf{r}_1| = |\mathbf{r}_2|$). Therefore, a *hidden* communication between the two detectors can be avoided by restricting the measurement to those events where $t_1 + 2\frac{|\mathbf{r}_1|}{c} > t_2 > t_1 + \frac{|\mathbf{r}_1|}{c}$. The two quantum paths contributing to a successful measurement are labeled red and black, respectively.

2. quantum path:

$$|e,-1;e,+1\rangle \otimes |0\rangle_1 |0\rangle_2 \;\to\; |e,-1;g,0\rangle \otimes |k\rangle_1 |0\rangle_2 \;\to\; |g,0;g,0\rangle \otimes |k\rangle_1 |k\rangle_2,$$

which differ by an optical phase $\delta_{21} = \delta_2 - \delta_1$. In case when the two detected photons do not share the same interval of existence we find it thus necessary to include the intermediate state of the atomic emitters which acts as a quantum memory storing the ambiguous which way information of the first detected photon in an entangled state where either of the two atoms could have emitted the photon. In this sense, the violation of the CHSH-type inequalities (6.4.6) might be interpreted as a proof of the entangled nature of the two correlated quantum paths, rather than the two photons contributing to the signal. We note that an experimental confirmation of these results might be carried out employing a time-loophole free test of the CHSH-type inequalities (see, e.g., [4]) by using the setup depicted in figure 6.2

Finally, let us point out, that there has been a similar debate about the actual nature of entanglement in case of the one-photon entangled states as discussed in chapter 1 (see, e.g., [31, 32]). In this context, again, it was found to be correct to speak of entangled quantum paths rather than of entangled particles.

6.5 Investigating spatial correlations for multiple emitters ($N > 2$)

In the remainder of this thesis we want to extend our investigations of spatial correlations to a system of N single-photon emitters. To simplify the discussion, we restrict ourselves to consider joint detection measurements only. In this case, a system of N emitters and two single-photon detectors gives rise to $N(N-1)$ possible quantum paths it might evolve along. We investigate the consequences of this increase in the number of the possibly occupied quantum paths by using again the CHSH-type inequalities.

6.5.1 Description of the physical system

We consider the setup shown in figure 6.3: N single-photon emitters are regularly arranged in a row serving as a source for single photons. Again, we use three-level systems with a V-configuration: for an even number of emitters, $N/2$ emitters are excited from the ground state along the transition $|g, 0\rangle \to |e, -1\rangle$ and $N/2$ emitters along $|g, 0\rangle \to |e, +1\rangle$ so that the initial state of the fully excited system can be written as

$$\prod_{n=1}^{N/2} |e_n, -1\rangle \otimes \prod_{n=1}^{N/2} |e_n, +1\rangle, \qquad (6.5.1)$$

whereas for an odd number of emitters, $(N-1)/2$ emitters are excited from the ground state along the transition $|g, 0\rangle \to |e, -1\rangle$ and $(N+1)/2$ emitters along $|g, 0\rangle \to |e, +1\rangle$,

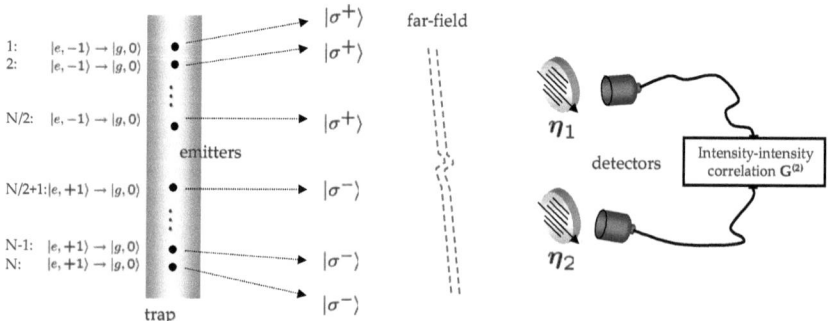

Figure 6.3: Setup used for measuring the intensity correlation function of second order for a source of N single-photon emitters.

and hence the initial state is given by

$$\prod_{n=1}^{\frac{N-1}{2}} |e_n, -1\rangle \otimes \prod_{n=1}^{\frac{N+1}{2}} |e_n, +1\rangle. \tag{6.5.2}$$

Again, two detectors are located in the far-field region of the emitters at $\mathbf{r}_1$ and $\mathbf{r}_2$ and equipped with polarizers oriented along $\boldsymbol{\eta}_1$ and $\boldsymbol{\eta}_2$, respectively. Recalling Eq. (6.2.2), we extend the detection operator for $N = 2$ emitters to the general case of an operator acting on the initial state (6.5.1) or (6.5.2) of N fully excited emitters given by

$$\hat{D}_N(\delta_j, \boldsymbol{\eta}_j) =$$
$$= \tfrac{\mathcal{E}_0}{\sqrt{2}} \left((\boldsymbol{\eta}_j \cdot \boldsymbol{\sigma}^+) \sum_{n=1}^{N/2} e^{in\delta_j} |g_n, 0\rangle\langle e_n, -1| + (\boldsymbol{\eta}_j \cdot \boldsymbol{\sigma}^-) \sum_{n=\frac{N+2}{2}}^{N} e^{in\delta_i} |g_n, 0\rangle\langle e_n, +1| \right)$$
$$= \tfrac{\mathcal{E}_0}{\sqrt{2}} \left(\sin\vartheta_j \sum_{n=1}^{N/2} e^{in\delta_j} |g_n, 0\rangle\langle e_n, -1| + \cos\vartheta_j \sum_{n=\frac{N+2}{2}}^{N} e^{in\delta_j} |g_n, 0\rangle\langle e_n, +1| \right), \tag{6.5.3}$$

where, again, we set the polarizer orientation $\boldsymbol{\eta}_j = \sin\vartheta_j \boldsymbol{\sigma}^+ + \cos\vartheta_j \boldsymbol{\sigma}^-$ so that $\mathcal{E}_j^2 := \mathcal{E}_0^2/(|(\boldsymbol{\eta}_j \cdot \boldsymbol{\epsilon}_1)|^2 + |(\boldsymbol{\eta}_j \cdot \boldsymbol{\epsilon}_2)|^2) = \mathcal{C}_0^2$ (c.f. Eq. (5.1.5)). In analogy, we obtain the detection operator for an odd number of N emitters

$$\hat{D}_N(\delta_j, \boldsymbol{\eta}_j) =$$
$$= \tfrac{\mathcal{E}_0}{\sqrt{2}} \left(\sin\vartheta_j \sum_{n=1}^{\frac{N-1}{2}} e^{in\delta_j} |g_n, 0\rangle\langle e_n, -1| + \cos\vartheta_j \sum_{n=\frac{N+1}{2}}^{N} e^{in\delta_j} |g_n, 0\rangle\langle e_n, +1| \right). \tag{6.5.4}$$

6.5.2 Intensity correlation signal of second order and its visibility for multiple emitters

With the expressions of Eq. (6.5.1) and (6.5.2) for the initial state and Eqs. (6.5.3) and (6.5.4) describing the detection operator, we can calculate the intensity correlation function of second order $G_N^{(2)}(\delta_1, \delta_2; \vartheta_1, \vartheta_2)$ for the setup shown in figure 6.3 assuming that the first two out of N scattered photons are detected (see section 6.5.3 for further information). Again, we fix the orientation of the polarization filters in front of the two detectors. Here, we consider two cases: first we look at the scenario where both polarizers are set identical to $\vartheta_1 = \vartheta_2 = \pi/4$ which corresponds to $\boldsymbol{\eta}_j = 1/\sqrt{2}(\boldsymbol{\sigma}^+ + \boldsymbol{\sigma}^-)$ (for $j = 1, 2$). In this case we find

$$G_N^{(2)}(\delta_1, \delta_2; \pi/4, \pi/4) = \frac{\mathcal{E}_0^4}{8} \left(1 + \frac{2}{N(N-1)} \sum_{n=1}^{N} (N-n) \cos(n(\delta_2 - \delta_1)) \right), \tag{6.5.5}$$

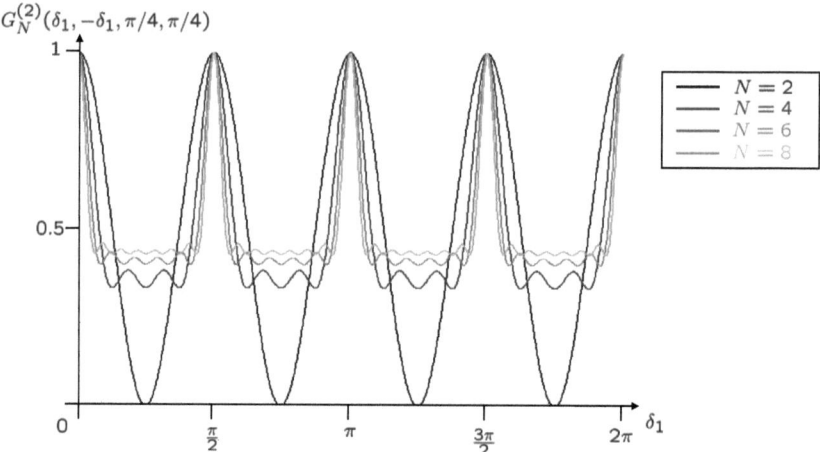

Figure 6.4: Plot of the intensity correlation function of second order $G_N^{(2)}(\delta_1, -\delta_2, \pi/4, \pi/4)$ (c.f. Eq. (6.5.5)) in arbitrary units. The plot illustrates the signal for different numbers of emitters N ($N = 2, 4, 6, 8$).

which holds for an arbitrary number of emitters, i.e., for even or odd N, and which is illustrated in figure 6.4. In the second scenario, we consider the two polarizers being set orthogonal at $\vartheta_1 = \pi/4$ and $\vartheta_2 = -\pi/4$ which translates into $\boldsymbol{\eta}_1 = 1/\sqrt{2}(\boldsymbol{\sigma}^+ + \boldsymbol{\sigma}^-)$ and $\boldsymbol{\eta}_2 = 1/\sqrt{2}(-\boldsymbol{\sigma}^+ + \boldsymbol{\sigma}^-)$. In this case we find for an even number of emitters N

$$G_N^{(2)}(\delta_1, \delta_2; \pi/4, 3\pi/4) = \qquad (6.5.6)$$
$$= \frac{\mathcal{E}_0^4}{8} \left(1 + \frac{2}{N(N-1)} \sum_{n=1}^{N/2} (N - 2n) \cos(n(\delta_2 - \delta_1)) \right.$$
$$\left. - \frac{2}{N(N-1)} \sum_{\alpha=1}^{N/2} \sum_{n=1}^{N} (\Theta(N - n - \alpha + 1) \cos(n(\delta_2 - \delta_1)) \Theta(n - \alpha + 1)) \right),$$

where we define the Heaviside step function $\Theta(x)$ as follows

$$\Theta(x) := \begin{cases} 0 & x \leq 0 \\ 1 & x > 0 \end{cases}. \qquad (6.5.7)$$

In analogy, we find for an odd number of emitters N

$$G_N^{(2)}(\delta_1, \delta_2; \pi/4, 3\pi/4) = \tag{6.5.8}$$

$$= \frac{\mathcal{E}_0^4}{8} \left(1 + \frac{2}{N(N-1)} \sum_{n=1}^{\frac{N-1}{2}} (N-2n) \cos(n(\delta_2 - \delta_1)) \right.$$

$$\left. - \frac{2}{N(N-1)} \sum_{\alpha=1}^{\frac{N-1}{2}} \sum_{n=1}^{N} (\Theta(N-n-\alpha+1) \cos(n(\delta_2-\delta_1)) \Theta(n-\alpha+1)) \right).$$

Finally, let us evaluate the visibility $\mathcal{V}_N$ of the intensity correlation signal of second order $G_N^{(2)}(\delta_1, \delta_2; \pi/4, \pi/4)$ as a function of the number of emitters N. For the expression of Eq. (6.5.5), which uses identical polarizer settings, we find the following dependency

$$\mathcal{V}_N := \frac{max[G_N^{(2)}] + min[G_N^{(2)}]}{max[G_N^{(2)}] - min[G_N^{(2)}]} = \frac{N}{3N-4}, \tag{6.5.9}$$

where $max[G_N^{(2)}]$ ($min[G_N^{(2)}]$) corresponds to the maximum (minimum) value of the function $G_N^{(2)} \equiv G_N^{(2)}(\delta_1, \delta_2; \pi/4, \pi/4)$. Eq. (6.5.9) shows that the visibility $\mathcal{V}_N$ can be assigned uniquely to the number of emitters N. We note, however, that $\mathcal{V}_N$ represents an ideal theoretical value only: in the foregoing sections we already discussed that experimental uncertainties and insufficiencies may influence the visibility (see section 4.2.4). From Eq. (6.5.9) we see that the maximal visibility $\mathcal{V}_N$ decreases for increasing numbers of emitters N, but also that it reaches a plateau of $\mathcal{V}_N = 1/3$ for $N \to \infty$.

6.5.3 Detection of two photons out of N scattered photons

In the foregoing section we derived the intensity correlation function of second order for N emitters by applying the detection operators of Eqs. (6.5.3) and (6.5.4) to the initially fully excited state of our system defined in Eqs. (6.5.1) and (6.5.2). Hereby, the calculations consider a scenario where it is assumed that the first two scattered photons are detected, excluding, e.g., the case where one of the emitters has scattered a photon which was not detected thereafter.

Let us briefly derive the consequences for our calculations if we do not consider that the initial state of the emitters is given by Eqs. (6.5.1) and (6.5.2), but allow for all possible subsets where only M atoms out of N possible emitters are excited ($M \leq N$) and contribute to the intensity correlation function of second order $G_{\binom{N}{M}}^{(2)}(\delta_1, \delta_2; \pi/4, \pi/4)$. In this case, we obtain the following relation [144]

$$G_{\binom{N}{M}}^{(2)}(\delta_1, \delta_2; \pi/4, \pi/4) = \frac{M(M-1)}{N(N-1)} G_N^{(2)}(\delta_1, \delta_2; \pi/4, \pi/4), \tag{6.5.10}$$

which demonstrates that the spatial dependence of the intensity correlation function $G^{(2)}_{\binom{N}{M}}$ is not changed as compared to $G^{(2)}_{\binom{N}{N}}$. However, we see from Eq. (6.5.10) that the probability for a successful measurement decreases with $\frac{M(M-1)}{N(N-1)}$, i.e., the fraction of atoms being initially in the ground state or in the excited state.

In an experiment trying to demonstrate a violation of CHSH-type inequalities for position correlations between two photons scattered by a source of multiple emitters $N > 2$, we thus have to make sure that the first two photons are detected. This can be implemented by detecting *all* N scattered photons but post-selecting the signal of the first two detected photons only.

6.5.4 Derivation of CHSH inequalities for spatial correlations and multiple emitters ($N > 2$)

In analogy to section 6.3.1, next, we derive CHSH-type inequalities which can be applied to a system of N emitters with V-configuration as depicted in figure 6.3. Again, we consider the mathematical inequalities (3.4.3). This time, we identify the following single detection probabilities

$$p^N(\delta_1, \vartheta_1, \lambda) = x, \ p^N(\delta'_1, \vartheta_1, \lambda) = x', \ p^N(\delta_1, \infty, \lambda) = X,$$
$$p^N(\delta_2, \vartheta_2, \lambda) = y, \ p^N(\delta'_2, \vartheta_2, \lambda) = y', \ p^N(\delta_2, \infty, \lambda) = Y, \qquad (6.5.11)$$

where ∞ indicates that the polarization filter is removed for the particular measurement[4] and the superscript N denotes the number of emitters used in the setup. In agreement with the requirements of an LHV theory, we postulate the joint probability to fulfill

$$p^N_{12}(\delta_1, \delta_2; \vartheta_1, \vartheta_2, \lambda) = p^N(\delta_1, \vartheta_1, \lambda) \cdot p^N(\delta_2, \vartheta_2, \lambda). \qquad (6.5.12)$$

Using Eqs. (6.5.11) and (6.5.12) together with the inequalities (3.4.3), we obtain, after multiplying the whole relation by $g(\lambda)$ and integrating over λ, the CHSH-type inequality

$$\begin{aligned}S^\star_N :=\ & p^N_{12}(\delta_1, \delta_2; \vartheta_1, \vartheta_2) - p^N_{12}(\delta_1, \delta'_2; \vartheta_1, \vartheta_2) + p^N_{12}(\delta'_1, \delta_2; \vartheta_1, \vartheta_2) \\ & + p^N_{12}(\delta'_1, \delta'_2; \vartheta_1, \vartheta_2) - p^N_{12}(\delta'_1, \delta_2; \vartheta_1, \infty) - p^N_{12}(\delta_1, \delta_2; \infty, \vartheta_2) \leq 0. \end{aligned} \qquad (6.5.13)$$

Here, we left out the normalization by a factor $p^N_{12}(\delta_1, \delta_2; \infty, \infty)$ and restricted ourselves to the inequality $S^\star_N \leq 0$: in difference to the case of $N = 2$ emitters (c.f. section 6.2), $p^N_{12}(\delta_1, \delta_2; \infty, \infty)$ for $N > 2$ is generally not a constant value but varies with δ_1 and δ_2. In

---
[4]Again, the constraint $X \geq x, x'$ ($Y \geq y, y'$) is guaranteed by the *no-enhancement* condition [59, 67, 68]: the detection probability with a polarization filter cannot exceed a measurement without a polarization filter.

order to avoid such an inappropriate normalization we rather consider the upper bound of the inequality (3.4.3) only, for which we can use an arbitrary normalization. For a better comparability, in the following, we thus choose a constant normalization by the factor of $p_{12}^2(\delta_1,\delta_2;\infty,\infty)$ independent of N so that the CHSH-type inequality reads

$$S_N := \left[p_{12}^N(\delta_1,\delta_2;\vartheta_1,\vartheta_2) - p_{12}^N(\delta_1,\delta_2';\vartheta_1,\vartheta_2) + p_{12}^N(\delta_1',\delta_2;\vartheta_1,\vartheta_2) + p_{12}^N(\delta_1',\delta_2';\vartheta_1,\vartheta_2)\right.$$
$$\left. - p_{12}^N(\delta_1',\delta_2;\vartheta_1,\infty) - p_{12}^N(\delta_1,\delta_2;\infty,\vartheta_2)\right]/p_{12}^2(\delta_1,\delta_2;\infty,\infty) \le 0. \quad (6.5.14)$$

6.5.5 Violating CHSH inequalities for multiple emitters ($N > 2$) by spatial correlations

In order to violate the inequality (6.5.14) maximally, it is advantageous, as in section 6.3.2, to adjust the polarization filters such that the detection efficiency of the experimental setup is optimized. In the following investigations we choose $\vartheta_1 = \vartheta_2 = \pi/4$ which yields the best results. With these values and using Eq. (2.2.9) we may calculate the joint detection probabilities needed in (6.5.14) as

$$p_{12}^N(\delta_1,\delta_2;\pi/4,\pi/4) = \frac{\mathcal{C}_0^2}{\mathcal{E}_0^4} G_N^2(\delta_1,\delta_2;\pi/4,\pi/4), \quad (6.5.15)$$

$$p_{12}^N(\delta_1,\delta_2;\pi/4,\infty) = \frac{\mathcal{C}_0^2}{\mathcal{E}_0^4} \left(G_N^2(\delta_1,\delta_2;\pi/4,\pi/4) + G_N^2(\delta_1,\delta_2;\pi/4,\pi/4+\pi/2)\right),$$

$$= \frac{\mathcal{C}_0^2}{\mathcal{E}_0^4} \left(G_N^2(\delta_1,\delta_2;\pi/4,\pi/4) + G_N^2(\delta_1,\delta_2;\pi/4,3\pi/4)\right), \quad (6.5.16)$$

$$p_{12}^2(\delta_1,\delta_2;\infty,\infty) = \mathcal{C}_0^2 \frac{1}{2}, \quad (6.5.17)$$

where we can make use of the expressions derived in Eqs. (6.5.5), (6.5.6) and (6.5.8). Whether or not a violation of the CHSH-type inequalities for $N > 2$ does occur can be verified by looking for the maxima of S_N as a function of δ_1, δ_2, δ_1' and δ_2'.

For the case of $N = 2$ emitters (c.f. section 6.2), the extrema of S_2 could be determined by using either of the two sets of values for $\delta_1, \delta_1', \delta_2, \delta_2'$ provided in Eqs. (3.4.8) and (3.4.9) which can be derived analytically (see, e.g., [68]). In this case we find

$$\delta_1 = \alpha_1\,2\pi,\quad \delta_2 = (\frac{1}{8}+\alpha_2)\,2\pi,\quad \delta_1' = (\frac{2}{8}+\alpha_3)\,\pi,\quad \delta_2' = (\frac{3}{8}+\alpha_4)\,\pi, \quad (6.5.18)$$

in agreement with the set of parameters given in Eq. (3.4.9).

In contrast, for $N > 2$, the joint detection probabilities present in S_N get more involved with increasing N (c.f. Eqs. (6.5.5), (6.5.6) and (6.5.8)). The values of $\delta_1, \delta_1', \delta_2, \delta_2'$ giving rise to maxima of S_N were thus determined numerically. This approach unveiled that the

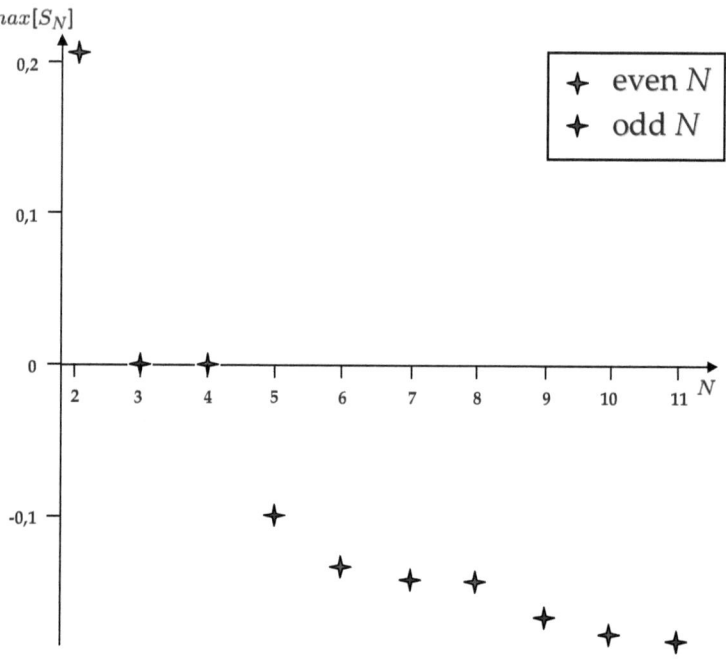

Figure 6.5: Maximum values of S_N in dependence of the number of emitters N: the numerical calculations show that the CHSH-type inequality (6.5.14) is violated only for $N = 2$, whereas for $N > 2$ the inequality holds. For $N = 3$ and $N = 4$ the maximum value is determined exactly as $S_{3,4} = 0$.

maxima of S_N can be obtained for any $N > 2$ simply by choosing

$$\delta_1 = \alpha_1\, 2\pi, \quad \delta_2 = \alpha_2\, 2\pi, \quad \delta'_1 = \alpha_3\, \pi, \quad \delta'_2 = \alpha_4\, \pi, \tag{6.5.19}$$

with α_j, $j = 1, ..., 4$, being arbitrary integers ($\alpha_3, \alpha_4 \neq 0$).

The results of our calculations for the maxima of S_N (for $N = 2, ..., 10$) are shown in figure 6.5. For $N = 2$ we obtain as before $S_2 = \frac{\sqrt{2}-1}{2}$ and for $N = 3, 4$ we find in both cases $S_3 = S_4 = 0$. For $N > 4$ the values of S_N cannot be cast into simple analytical expressions but are displayed in the plot: we see that the behavior is slightly different for even N (red line) and for odd N (blue line), however, we find that a violation of the CHSH-type inequality appears only for the case $N = 2$.

6.5.6 Conclusion

We see from figure 6.5 that the CHSH-type inequalities (6.5.14) cannot be violated by the setup shown in figure 6.3 for $N > 2$ emitters. This result is in agreement with some of our earlier investigations (c.f. section 6.3): for $N = 2$ emitters we found that we have to achieve a visibility of at least 71% if we want to demonstrate a violation of CHSH-type inequalities in a real experiment.

While the joint detection probability for our setup in case of $N = 2$ shows a modulation with a theoretical visibility of $\mathcal{V}_2 = 100\%$, we know from Eq. (6.5.9) that $\mathcal{V}_N$ drops rapidly with increasing $N > 2$. The corresponding visibility $\mathcal{V}_N$ is illustrated in figure 6.6. As it is shown, already for the case of $N = 3$ emitters we obtain a maximal visibility of $\mathcal{V}_3 = 60\%$ only, i.e., the maximally achievable visibility drops below the required value of 71%. Triggered by these results, we will consider a different inequality which is more suitable for our system in the next section. As it turns out the latter is able to prove that the spatial correlations of spontaneously emitted photons exhibit a non-classical nature even in case of $N > 2$ emitters.

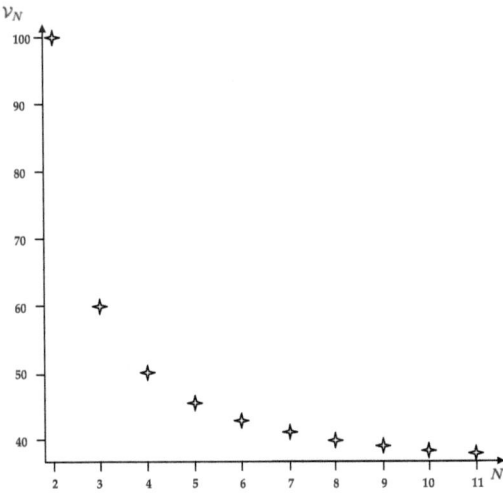

Figure 6.6: Plot of the visibility $\mathcal{V}_N$ (c.f. Eq. (6.5.9)) of the modulated intensity correlation signal of second order in dependence of the number of emitters N; the visibility follows the function $\mathcal{V}_N = \frac{N}{3N-4}$ which reaches 50% for $N = 4$ and 33% for $\mathcal{V}_N \rightarrow \infty$, respectively.

6.6 A more suitable inequality for multiple emitters

In the final part of this thesis, we consider another inequality which turns out to be more suitable for our system introduced in section 6.5 and illustrated in figure 6.3. In order to achieve this, we rederive a CHSH-type inequality, however, starting with a different mathematical inequality based on the so-called Bell Wigner-inequality (see, e.g., [145, 146]).

6.6.1 Derivation of a homogeneous Bell-Wigner (HBW) inequality

The Bell-Wigner inequality can be written as [146]

$$0 \leq x_1 - x_1 x_2 - x_1 x_3 + x_2 x_3, \qquad (6.6.1)$$

which holds for the constraints that $0 \leq x_1, x_2, x_3 \leq 1$; for a proof of this inequality we refer to [145]. Here, identifying x_j with single detection probabilities ($j = 1, 2, 3$) and following the notation used in this thesis, we could speak of a Bell-type inequality since it considers single and joint detection probabilities. However, as motivated before, the experimental requirements can be eased if the inequality under investigation involves only detection probabilities of the same order. Hence, our goal is to derive a novel CHSH-type inequality on the basis of the Bell-Wigner inequality which considers joint detection probabilities only, being subject to the same overall success probability. Our proposal for a new inequality reads

$$0 \leq x_1 x_4 - x_1 x_2 - x_1 x_3 + x_2 x_3, \qquad (6.6.2)$$

and holds for the constraints $0 \leq x_1, x_2, x_3 \leq x_4 \leq 1$. The proof of (6.6.2) is provided in appendix C.

In analogy to the foregoing section, we consider the setup with an even (odd) number of emitters N as displayed in figure 6.3, i.e., where $N/2$ (($N-1$)/2) emitters scatter $\boldsymbol{\sigma}^-$ polarized light and another $N/2$ (($N+1$)/2) emitters scatter $\boldsymbol{\sigma}^+$ polarized photons. A photon event registered at the jth detector is characterized by two parameters, the position $\mathbf{r}_j$ giving rise to an optical phase δ_j and the orientation of the jth polarizer which we choose to be oriented along $\boldsymbol{\eta}_j = 1/\sqrt{2}(\boldsymbol{\sigma}^- + \boldsymbol{\sigma}^+)$ ($j = 1, 2$) corresponding to $\vartheta_2 = \vartheta_1 = \pi/4$. The latter optimizes the overall success probability. The detection operators for this system were derived in Eqs. (6.5.3) and (6.5.4). In addition, we make use of the results found for $G_N^{(2)}(\delta_1, \delta_2; \pi/4, \pi/4)$ and $G_N^{(2)}(\delta_1, \delta_2; \pi/4, 3\pi/4)$ in Eqs. (6.5.5), (6.5.6) and (6.5.8), respectively.

Here, we identify the parameters of (6.6.2) with the detection probabilities

$$p^N(\delta_1, \vartheta_1, \lambda) = x_1, \; p^N(\delta_2, \vartheta_1, \lambda) = x_2,$$
$$p^N(\delta_3, \vartheta_2, \lambda) = x_3, \; p^N(\delta_4, \infty, \lambda) = x_4, \quad (6.6.3)$$

where ∞ indicates that the polarization filter is removed for the particular measurement[5] and, following the usual assumptions of an LHV theory, we define the joint detection probability exactly as in Eq. (6.5.12). Using Eqs. (6.6.3) and (6.5.12) in the new inequality (6.6.2), we obtain, after multiplying the whole relation by $g(\lambda)$ and integrating over λ,

$$\begin{aligned} S_N^\star &:= p_{12}^N(\delta_1, \delta_4; \vartheta_1, \infty) - p_{12}^N(\delta_1, \delta_2; \vartheta_1, \vartheta_1) \\ &\quad - p_{12}^N(\delta_1, \delta_3; \vartheta_1, \vartheta_2) + p_{12}^N(\delta_2, \delta_3; \vartheta_1, \vartheta_2) \geq 0. \end{aligned} \quad (6.6.4)$$

For a better comparability with the results obtained so far and in analogy to the foregoing section, we introduce a normalization by the factor $p_{12}^2(\delta_1, \delta_2; \infty, \infty)$ (c.f. Eq. 6.5.17) independent of N so that the inequality finally reads

$$\begin{aligned} S_N &:= \left[p_{12}^N(\delta_1, \delta_4; \vartheta_1, \infty) - p_{12}^N(\delta_1, \delta_2; \vartheta_1, \vartheta_1) \right. \\ &\quad \left. - p_{12}^N(\delta_1, \delta_3; \vartheta_1, \vartheta_2) + p_{12}^N(\delta_2, \delta_3; \vartheta_1, \vartheta_2) \right] / p_{12}^2(\delta_1, \delta_2; \infty, \infty) \geq 0. \end{aligned} \quad (6.6.5)$$

In the remainder of this thesis, we refer to the inequality (6.6.5) as *homogeneous Bell-Wigner* (HBW) inequality.

6.6.2 Violation of the HBW inequality for $N \geq 2$ emitters by spatial correlations

Here we will test the HBW inequality (6.6.5) for $N \geq 2$ emitters by evaluating the minimum values of S_N. Therefore, we identify the following joint detection probabilities by applying Eq. (2.2.9)

$$p_{12}^N(\delta_1, \delta_2; \pi/4, \pi/4) = \frac{C_0^2}{\mathcal{E}_0^4} G_N^{(2)}(\delta_1, \delta_2, \pi/4, \pi/4), \quad (6.6.6)$$

$$p_{12}^N(\delta_1, \delta_2; \pi/4, \infty) = \frac{C_0^2}{\mathcal{E}_0^4} \left(G_N^{(2)}(\delta_1, \delta_2, \pi/4, \pi/4) + G_N^{(2)}(\delta_1, \delta_2, \pi/4, 3\pi/4) \right), \quad (6.6.7)$$

$$p_{12}^2(\delta_1, \delta_2; \infty, \infty) = C_0^2 \frac{1}{2}, \quad (6.6.8)$$

---

[5]Again, the constraint $X \geq x, x'$ ($Y \geq y, y'$) is guaranteed by the *no-enhancement* condition [59, 67, 68]: the detection probability with a polarization filter cannot exceed the measurement without a polarization filter.

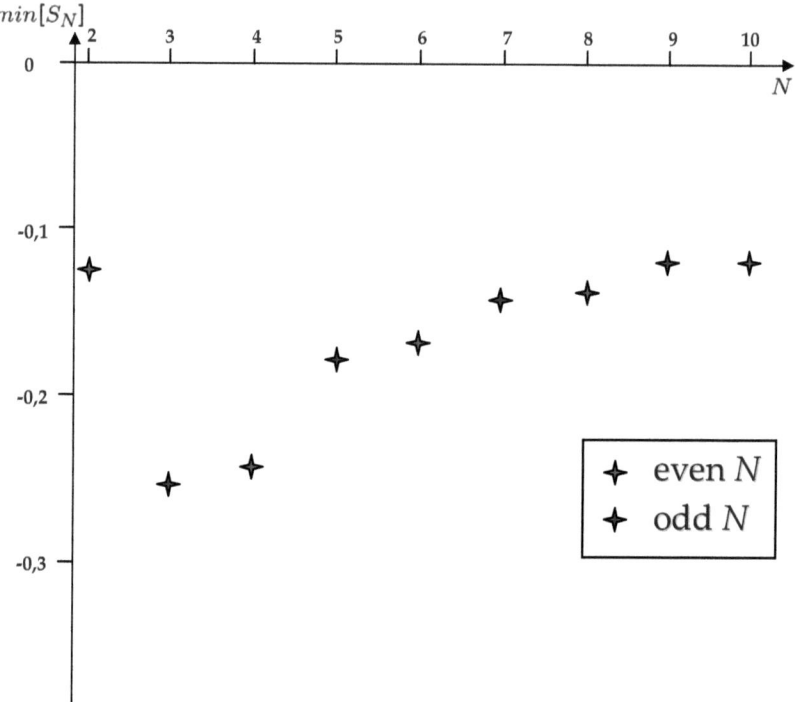

Figure 6.7: Minima values of S_N (c.f. Eq. (6.6.5)) for $N = 2, ..., 10$; the plot illustrates a steady violation of the HBW inequality of Eq. (6.6.5) with a decreasing amplitude. For $N = 2$ the minima value of S_2 is much greater than expected - it is rather comparable with the minima value of S_{10}.

where the last equation resembles Eq. (6.5.17).

In principle, we can use Eqs. (6.5.5), (6.5.6) and (6.5.8) in the HBW inequality (6.6.5) and determine the minimum values of S_N. However, as the analyses get involved and analytically impracticable, here, we only provide numerical results, where we scanned through the complete parameter space of $\delta_1, \delta_2, \delta_3, \delta_4$ looking for the minima values of S_N for each N separately.

The results for the minima of S_N are shown in figure 6.7: the plot illustrates the minima of S_N, i.e., $min[S_N]$, for $N = 2, ..., 10$. As can be seen from the plot, it suggests that a violation of the HBW inequalities (6.6.5) can be obtained for any N. For $N > 2$, the values of $min[S_N]$ decrease monotonously approaching zero for $N \to \infty$. For $N = 3$ we obtain the lowest value of $min[S_3] \approx -0.254$. The smallest amplitude shown in the plot

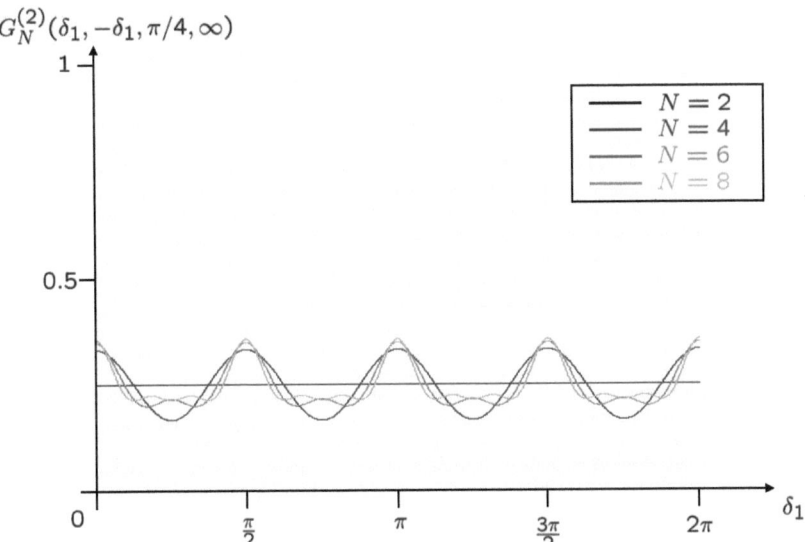

Figure 6.8: Plot of the intensity correlation function of second order $G_N^{(2)}(\delta_1, -\delta_1, \pi/4, \infty) := G_N^{(2)}(\delta_1, \delta_1, \pi/4, \pi/4) + G_N^{(2)}(\delta_1, \delta_1, \pi/4, 3\pi/4)$ in arbitrary units (c.f. Eq. (6.6.7)). The plot illustrates the signal for different numbers of emitters N ($N = 2, 4, 6, 8$).

is obtained for $N = 10$ corresponding to $min[S_{10}] \approx -0.118$.

We remark that for $N = 2$ we obtain $min[S_2] = -0.125$ which sticks out of the overall behavior. We explain this outlier by the fact that S_2 depends only on three of the four parameters $\delta_1, \delta_2, \delta_3, \delta_4$ since, due to destructive interference, we have $p_{12}^2(\delta_1, \delta_4; \vartheta_1, \infty) = \frac{C_0^2}{4}$, i.e., a constant independent of δ_1 and δ_4. In contrast, for S_N with $N > 2$ the term $p_{12}^N(\delta_1, \delta_4; \vartheta_1, \infty)$ is no longer a constant value and thus can be employed to shift the value of $min[S_N]$ towards smaller values. Figure 6.8 illustrates the intensity correlation function $G_N^{(2)}(\delta_1, -\delta_1, \pi/4, \infty) := G_N^{(2)}(\delta_1, \delta_1, \pi/4, \pi/4) + G_N^{(2)}(\delta_1, \delta_1, \pi/4, 3\pi/4)$ for $N = 2, 4, 6, 8$.

6.6.3 Conclusion: violation of the HBW inequality and the visibility of the correlated signal

Let us consider the theoretically attainable visibility of the correlated signal measured: from Eq. (6.5.9) (c.f figure 6.6) we found that the visibility of the intensity correlation function of second order $G_N^{(2)}(\delta_1, \delta_2; \pi/4, \pi/4)$ follows the function $\mathcal{V}_N = \frac{N}{3N-4}$ which reaches 50% for $N = 4$ and 33% for $N \to \infty$, respectively. However, from figure 6.7 we see that the HBW inequalities remain continuously violated when increasing the number of emitters N: only for the limiting case $N \to \infty$ the value of $min[S_N]$ converts to zero where no violation of the HBW inequalities is obtained, while the visibility reaches 33%. In other words, our results proof that the HBW inequality (6.6.5) can be violated by an intensity correlation signal of second order if exhibiting a visibility $\mathcal{V}_N > 33\%$.

A similar behavior is known in the literature in the context of Werner-states [147]: in 1989 Werner constructed a LHV model, where he introduced a mixture of an entangled pure state $|\Psi\rangle$ with noise parameterized by the positive parameter p

$$\rho_p^W = p|\Psi\rangle\langle\Psi| + \frac{1-p}{4}\mathbb{1}. \qquad (6.6.9)$$

Werner showed that these states are separable for $p \leq 1/3$, admit his LHV model for projective measurements for $p \leq 1/2$ and violate the CHSH inequality (c.f. (6.5.14)) only for $p > 1/\sqrt{2}$ [147]. Thereby, for $1/3 < p \leq 1/2$, he proved that entanglement does not necessarily lead to non-local behavior incompatible with any LHV theory. Thereafter a number of authors continued to investigate the non-local behavior of the Werner-state [148–153]. Only recently Wildfeuer and Dowling showed that Werner-like states can violate Bell-type inequalities similar to the inequality (6.6.1) for the entire range $p > 1/3$, i.e., where the state of Eq. (6.6.9) cannot be separated, thus proving its entanglement [153]. The HBW inequality (6.6.5) extended these inequalities naturally: in analogy to the CHSH inequalities, it consists only of joint detection probabilities being subject to the same overall success probability, which is thus less demanding for an experimental test.

In comparison, for our scheme, we found that the theoretically attainable visibility $\mathcal{V}_N$ of the signal of our correlation measurement determines whether or not a violation of, e.g., CHSH-type inequalities can be obtained: in particular, for the CHSH-type inequality (6.5.14) we proved that a visibility $\mathcal{V}_2 > 1/\sqrt{2} \approx 71\%$ of the correlated signal must be achieved in order to obtain a violation. Here, in the present section, we found that the HBW inequality (6.6.5) can be violated by a correlated two-photon signal with a visibility $\mathcal{V}_N > 1/3 \approx 33\%$. On the other hand, when using classical light sources obeying thermal statistics, it is well known that intensity correlations of second order cannot exceed a visibility of 33% [92]. Any correlation signal that reveals a higher visibility is thus a hint for a *non-classical* light field. Obviously, this limit is in agreement with our results.

One can argue whether the visibility of the intensity correlation signal of second order provides a good parameter to estimate the classical or entangled character of the two-photon signal measured. In the light of the foregoing analyses, we conclude with the striking comparison between our results on a violation of the HBW inequality and the results of Wildfeuer and Dowling, i.e., with the comparison between p and $\mathcal{V}_N$.

Chapter 7

Conclusions

We conclude this thesis by summarizing the different research topics which were covered and the most important results which were derived. Though the thesis treated three conceptually different topics in chapters 4, 5 and 6, namely quantum imaging, quantum state engineering and testing quantum correlations by violating Bell inequalities, all our investigations are based commonly on the very basic setup of single-photon emitters in a particular geometric arrangement wherein we performed correlation measurements on the fluorescence light in the far-field region of the emitters as introduced in chapter 2. The results found should be attainable in an experiment using trapped ions as an optimized source of single photons while providing technical feasibility.

In chapter 4, we examined the fluorescence light scattered by a source of single-photon emitters arranged in a geometric array. Performing intensity measurements we showed how a correlated signal can be utilized to enhance imaging characteristics. In section 4.2, we considered the most simple scenario of a far-field imaging scheme where the source itself, i.e., the atomic emitters, is subject to the imaging process. We demonstrated that the intensity correlation signal of Nth order as measured in the fluorescence light of N emitters allows to image the latter with a resolution N-times better than in a classical analog scenario [19]. In section 4.3, we extended the setup under consideration to allow also for imaging of real physical objects. Employing classical diffraction theory together with the quantum nature of the electromagnetic field generated by N single-photon emitters, we proved that the diffraction pattern of an object can be restored with a resolution N-times better (for $N = 2, 4$) than in a classical analog scenario [23]. Hereby, keeping a contrast of 100%, our far-field imaging scheme provides the first experimental setup which might image almost arbitrary objects with sub-classical resolution while using linear optical tools only which guarantees technical feasibility and scalability.

In chapter 5, we focused on the quantum state of the single-photon emitters after performing correlation measurements in their far-field region. We introduced three-level emitters with Λ-level structure providing two long-lived ground states acting as qubit in our scheme. The particular ground states can be projected at will by employing a polarization sensitive detection scheme. In combination with optical phase differences which result from distinct possible quantum paths the single photons can evolve along from the emitters to the detectors, we showed how to engineer certain multi-qubit quantum states. In section 5.2, we demonstrated how these degrees of freedom can be used to generate any symmetric Dicke state for an arbitrary number of emitters, including the whole multi-partite entangled class of W-states [21]. In section 5.3, we modified the basic setup of chapter 2 insofar as the far-field detection scheme is substituted by a detection scheme based on optical fibers. Hereby, each of the possible pathways a single photon can evolve along is represented by an optical fiber leading from any of the N emitters to the N detectors. In particular, each optical fiber can define an arbitrary unique optical phase the photon accumulates before being detected. We demonstrated how this increased parameter space can be fruitfully exploited to generate any of the total angular momentum eigenstates of an N-qubit system [24]. In section 5.4, we finally utilized the full parameter space provided by a polarization sensitive detection scheme: allowing for projecting arbitrary linear polarizations, we showed how to generate symmetric quantum states belonging to any of the three tripartite entanglement classes, i.e., the class of GHZ-states, of W-states and of separable states. Remarkably, switching between these classes is allowed simply by tuning the orientation of the polarizers [22].

Finally, in chapter 6, we investigated the quantum nature of the correlated fluorescence signal of the setup introduced in chapter 2 which was utilized in all our analyses throughout this thesis. We derived CHSH inequalities for our system in order to prove the entangled nature of the photons in the fluorescence signal by utilizing polarization degrees of freedom. There, we found that a violation of the CHSH inequalities can be achieved in a realistic experiment. In addition, by looking at the dynamics of our system we derived an interesting interpretation of the violation found for the CHSH inequalities for the present system: the two emitted photons do not have to share the same interval of existence despite the fact that their correlations might contribute to the violation of the inequalities. Our results thus strengthen a recent development in the research field of quantum information science where *we must not view non-locality as pertaining to particles themselves, but see it instead as a property of quantum fields whose significance is, therefore, more fundamental than that of particles* [154]. At the end of chapter 6, we described how CHSH inequalities can be applied to a system of $N > 2$ single-photon emitters. It turned out that as the visibility of the correlation signal drops rapidly with N no such violation can be obtained for $N > 2$ since a visibility greater than 71% is required. However, deriving a more suitable homogenous Bell-Wigner inequality, we showed that it is possible to obtain a violation of this inequality for any number of emitters N and in particular for a visibility greater than 33%.

Appendix A

Expectation values for multi-time intensity correlations

Recalling the expressions of Eqs. (2.2.1) and (2.2.7), the intensity correlation function of Nth order where N photons scattered by N emitters are registered at times $t_1, ..., t_N$ and at positions $\mathbf{r}_1, ..., \mathbf{r}_N$ contains expectation values of the form

$$\left\langle \prod_{n=1}^{N} \hat{S}_n^+(t_n) \prod_{n=1}^{N} \hat{S}_{N-n+1}^-(t_{N-n+1}) \right\rangle, \tag{A.0.1}$$

where a time ordering is assumed, i.e., $t_1 < t_2 < ... < t_N$. Using the master equation in the limit $d \gg \lambda$ (c.f. Eq. (2.1.8))

$$\frac{\partial \rho}{\partial t} = -i\omega \left[\sum_n^N \hat{S}_n^z, \rho \right] - \gamma \sum_n^N \left(\hat{S}_n^+ \hat{S}_n^- \rho - 2\hat{S}_n^- \rho \hat{S}_n^+ + \rho \hat{S}_n^+ \hat{S}_n^- \right). \tag{A.0.2}$$

We have solved the equations of motion for some of these terms (c.f. Eqs. (2.1.11) and (2.1.13))

$$\left\langle \hat{S}_n^\pm(t) \right\rangle = e^{(\pm i\omega - \gamma)t} \left\langle \hat{S}_n^\pm(0) \right\rangle, \tag{A.0.3}$$

$$\left\langle \prod_n^m \hat{S}_n^+(t) \prod_n^m \hat{S}_{m-n+1}^-(t) \right\rangle = e^{-2m\gamma t} \left\langle \prod_n^m \hat{S}_n^+(0) \prod_n^m \hat{S}_{m-n+1}^-(0) \right\rangle, \tag{A.0.4}$$

which hold for any number $m \leq N$. However, in order to display the time dependence of the atomic operators acting at N different times (c.f. (A.0.1)) needed to derive the intensity correlation function of Nth order we have to apply the quantum regression theorem: the regression theorem states that if the one-time expectation values fulfill the relation (c.f. [55])

$$\left\langle \hat{Q}_n(t_2) \right\rangle = f(t_2 - t_1)\left\langle \hat{Q}_n(t_1) \right\rangle \quad \text{for } t_2 > t_1, \tag{A.0.5}$$

then the two-time correlations obey

$$\left\langle \hat{Q}_m(t_1)\hat{Q}_n(t_2) \right\rangle = f(t_2 - t_1)\left\langle \hat{Q}_m(t_1)\hat{Q}_n(t_1) \right\rangle \quad \text{for } t_2 > t_1. \tag{A.0.6}$$

By successively applying the quantum regression theorem to the term (A.0.1) we thus find

$$\left\langle \prod_{n=1}^{N} \hat{S}_n^+(t_n) \prod_{n=1}^{N} \hat{S}_{N-n+1}^-(t_{N-n+1}) \right\rangle = \tag{A.0.7}$$

$$e^{(i\omega-\gamma)(t_N - t_{N-1})} \left\langle \prod_{n=1}^{N-1} \hat{S}_n^+(t_n)\hat{S}_N^+(t_{N-1}) \prod_{n=1}^{N} \hat{S}_{N-n+1}^-(t_{N-n+1}) \right\rangle =$$

$$e^{-2\gamma(t_N - t_{N-1})} \left\langle \prod_{n=1}^{N-1} \hat{S}_n^+(t_n)\hat{S}_N^+(t_{N-1})\hat{S}_N^-(t_{N-1}) \prod_{n=2}^{N} \hat{S}_{N-n+1}^-(t_{N-n+1}) \right\rangle =$$

$$\prod_{n=1}^{N-1} e^{-2n\gamma(t_{N-n+1} - t_{N-n})} \left\langle \prod_{n}^{N} \hat{S}_n^+(t_1) \prod_{n}^{N} \hat{S}_{N-n+1}^-(t_1) \right\rangle =$$

$$\prod_{n=1}^{N} e^{-2n\gamma(t_{N-n+1} - t_{N-n})} \left\langle \prod_{n}^{N} \hat{S}_n^+(0) \prod_{n}^{N} \hat{S}_{N-n+1}^-(0) \right\rangle, \quad \text{for } t_0 \equiv 0.$$

Please note that this outcome for the time dependence of the expectation value does not change if the operators $\hat{S}_n^\pm$ and the times t_n do not carry the same subscript. Therefore, with the expression of Eq. (A.0.7), we find from Eqs. (2.2.1) and (2.2.7)

$$G^{(N)}(\mathbf{r}_1, t_1; ...; \mathbf{r}_N, t_n) = \prod_{n=1}^{N} e^{-2n\gamma(t_{N-n+1} - t_{N-n})} G^{(N)}(\mathbf{r}_1, ..., \mathbf{r}_N), \tag{A.0.8}$$

i.e., the time dependence of the intensity correlation function of Nth order factorizes and thus justifies the simplified treatment employed throughout this thesis by using the basic timed independent detection operator introduced in Eq. (2.2.13) or (2.2.14), respectively.

Appendix B

Derivation of Eq. (4.3.2)

From classical diffraction optics (see, e.g., [98]) we know that the electric field amplitude $E(\mathbf{r})$ is diffracted at an aperture $\mathcal{A}$ placed between the source and the detection plane. In the far-field region of the aperture, the *Fresnel-Kirchhoff* integral formula for the diffracted field at $\mathbf{r}_j$ produced by a point source of unit strength at $\mathbf{R}_n$ can be written as

$$U(\mathbf{r}_j, \mathbf{R}_n) = -\frac{iA}{2\lambda} \iint_{\mathcal{A}} \frac{e^{ik|\mathbf{R}_n - \mathbf{r}_0|}}{|\mathbf{R}_n - \mathbf{r}_0|} \frac{e^{ik|\mathbf{r}_0 - \mathbf{r}_j|}}{|\mathbf{r}_0 - \mathbf{r}_j|} \left[\cos(\mathbf{r}_0, \mathbf{R}_n) - \cos(\mathbf{r}_0, \mathbf{r}_j)\right] dS, \quad (B.0.1)$$

where A is the amplitude of the electric field at unit distance from the source [98], $k = \frac{2\pi}{\lambda}$ denotes the wavenumber and $\mathbf{r}_0$ is a vector in the plane of the aperture (see figure 4.6). Here, the propagation of light from the source to the aperture and from the latter to the detector in the Fourier plane is described by the two Green's functions in the integrand. Though the integral of Eq. (B.0.1) describes the full diffraction problem in a general manner it is not solvable analytically. There are two approximations commonly used in order to simplify the problem, namely the *Fraunhofer-* and the *Fresnel*-approximation. Both methods provide good agreement with experimental data if the plane of observation is in the far-field region of the aperture and if the paraxial approximation holds, i.e., if source, aperture and detector are not too far off from the optical axes.

The far-field approximations can be summarized as

- $|\mathbf{r}_0 - \mathbf{r}_j| \approx |r_{0_z} - r_{j_z}| := r_z$ and $|\mathbf{R}_n - \mathbf{r}_0| \approx |R_{n_z} - r_{0_z}| := R_z$,
- $[\cos(\mathbf{r}_0, \mathbf{R}_n) - \cos(\mathbf{r}_0, \mathbf{r}_j)] \approx 2$.

Using these legitimate approximations the *Fresnel-Kirchhoff* integral reduces to the so called *Fresnel-integral* of Eq. (4.3.2) which reads

$$U(\mathbf{r}_j, \mathbf{R}_n) = -\frac{iA}{\lambda} \frac{e^{ikR_z} e^{ikr_z}}{R_z r_z} \iint_{\mathcal{A}} e^{i\frac{k}{2}\frac{|\boldsymbol{\rho}_n - \boldsymbol{\rho}_0|^2}{R_z}} e^{i\frac{k}{2}\frac{|\boldsymbol{\rho}_0 - \boldsymbol{\rho}_j|^2}{r_z}} dS(\boldsymbol{\rho}_0), \quad (B.0.2)$$

where $\boldsymbol{\varrho}_n$, $\boldsymbol{\rho}_0$ and $\boldsymbol{\rho}_j$ are vectors consisting of the x- and y-components of $\mathbf{R}_n$, $\mathbf{r}_0$ and $\mathbf{r}_j$, respectively (with $n, j = 1, 2$).

Though, the *Fresnel-integral* can be solved numerically, we make further use of the *Fraunhofer* approximation

$$R_z, r_z \gg \frac{k}{2} |\boldsymbol{\rho}_0|^2, \tag{B.0.3}$$

which leads to the analytically solvable *Fraunhofer-integral* which we applied to derive Eq. (4.3.4). For this case, the varying quadratic terms in the exponents of the integrand can be approximated by

$$\begin{aligned} \frac{ik}{2} \frac{|\boldsymbol{\varrho}_n - \boldsymbol{\rho}_0|^2}{R_z} &= \frac{ik}{2R_z} \left[|\boldsymbol{\varrho}_n|^2 - 2(R_{n_x} r_{0_x} + R_{n_y} r_{0_y}) + |\boldsymbol{\rho}_0|^2 \right] \\ &\approx \frac{ik}{2R_z} \left[|\boldsymbol{\varrho}_n|^2 - 2(R_{n_x} r_{0_x} + R_{n_y} r_{0_y}) \right], \end{aligned} \tag{B.0.4}$$

and

$$\begin{aligned} \frac{ik}{2} \frac{|\boldsymbol{\rho}_0 - \boldsymbol{\rho}_i|^2}{r_z} &= \frac{ik}{2r_z} \left[|\boldsymbol{\rho}_0|^2 - 2(r_{0_x} r_{i_x} + r_{0_y} r_{i_y}) + |\boldsymbol{\rho}_i|^2 \right] \\ &\approx \frac{ik}{2r_z} \left[-2(r_{0_x} r_{i_x} + r_{0_y} r_{i_y}) + |\boldsymbol{\rho}_i|^2 \right]. \end{aligned} \tag{B.0.5}$$

Appendix C

Proof of the inequality (6.6.2)

Here, we prove the inequality (6.6.2), since it differs slightly from the Bell-Wigner inequality (6.6.1) (see, e.g., [145]). The Bell-Wigner inequality usually reads

$$0 \leq x_1 - x_1 x_2 - x_1 x_3 + x_2 x_3, \tag{C.0.1}$$

where $0 \leq x_1, x_2, x_3 \leq 1$ has to be fulfilled.

As explained in section 6.5, for experimental reasons it is advantageous to use the inequality (6.6.2)

$$0 \leq x_1 x_4 - x_1 x_2 - x_1 x_3 + x_2 x_3, \tag{C.0.2}$$

consisting of products of the form $x_i x_j$ ($i, j = 1, 2, 3, 4$) only. For the proof of (6.6.2) it is essential that $1 \geq x_4 \geq x_1, x_2, x_3 \geq 0$. This enables us later on to use the *no-enhancement* assumption for our system (c.f. [59]).

We consider two cases:

First, $x_2 \geq x_1$ is assumed. In this case we rewrite the inequality (C.0.2) as

$$0 \leq x_1 (x_4 - x_2) + x_3 (x_2 - x_1), \tag{C.0.3}$$

which is valid since both brackets are positive or equal zero due to $x_2 \geq x_1$ and $x_4 \geq x_2$ (we also note that $x_1, x_2, x_3, x_4 \geq 0$).

Second, we assume $x_1 > x_2$. Here, we make a further case differentiation: let us assume $x_1 \geq x_3$ and rewrite the inequality (C.0.2) as

$$0 \leq x_1 (x_4 - x_2) - x_3 (x_1 - x_2). \tag{C.0.4}$$

This inequality is valid since the first bracket is bigger or equals the second due to $x_4 \geq x_1$

and since $x_1 \geq x_3$. In contrast, if we assume $x_3 > x_1$, we rewrite the inequality (C.0.2) as

$$x_1(x_4 - x_3) - x_2(x_1 - x_3) \stackrel{x_3 > x_1}{>} x_1(x_4 - x_3) \geq 0, \qquad (C.0.5)$$

where the last inequality holds due to $x_4 \geq x_3$.

Bibliography

[1] A. Einstein, B. Podolsky, and N. Rosen, Phys. Rev. **47**, 777 (1935).

[2] E. Schrödinger, Naturwissenschaften, **23** 807; *ibid* 823; *ibid* 844 (1935).

[3] Albert Einstein to Max Born, 1947, M. Born, *The Born-Einstein Letters*, translated by Irene Born, Macmillan, London (1971).

[4] A. Aspect, J. Dalibard, and G. Roger, Phys. Rev. Lett. **49**, 1804 (1982).

[5] P. G. Kwiat *et al.*, Phys. Rev. Lett. **75**, 4337 (1995).

[6] S. Osnaghi *et al.*, Phys. Rev. Lett. **87**, 037902 (2001).

[7] O. Mandel *et al.*, Nature **425**, 937 (2003).

[8] D. Leibfried *et al.*, Nature (London), **438**, 639 (2005).

[9] C. A. Häffner *et al.*, Nature (London) **438**, 643 (2005).

[10] T. Wilk *et al.*, Science **317**, 488 (2007).

[11] E. S. Polzik, Phys. Rev. A **59**, 4202 (1999).

[12] C. Cabrillo *et al.*, Phys. Rev. A **59**, 1025 (1999).

[13] S. Bose *et al.*, Phys. Rev. Lett. **83**, 5158 (1999).

[14] L.-M. Duan *et al.*, Nature **414**, 413 (2001).

[15] J. Hong and H.-W. Lee, Phys. Rev. Lett. **89**, 237901 (2002).

[16] X.-L. Feng *et al.*, Phys. Rev. Lett. **90**, 217902 (2003).

[17] L.-M. Duan and H. J. Kimble, Phys. Rev. Lett. **90**, 253601 (2003).

[18] C. Simon and W. T. M. Irvine, Phys. Rev. Lett. **91**, 110405 (2003).

[19] C. Thiel, T. Bastin, J. Martin, E. Solano, J. von Zanthier, and G. S. Agarwal, Phys. Rev. Lett. **99**, 133603 (2007).

[20] D. L. Moehring et al., Nature (London) **449**, 68 (2007).

[21] C. Thiel, J. von Zanthier, T. Bastin, E. Solano, and G. S. Agarwal, Phys. Rev. Lett. **99**, 193602 (2007).

[22] T. Bastin et al., Phys. Rev. Lett. **102**, 053601 (2009).

[23] C. Thiel et al., accepted in Phys. Rev. A; quant-ph/08051831 (2009).

[24] A. Maser et al., Phys. Rev. A, **79**, 033833 (2009).

[25] U. Schilling et al., submitted to Physical Review Letters; quant-ph/09012592 (2009).

[26] T. Young, Phil. Trans. R. Soc. Lond. **94**, 1 (1804).

[27] A. Einstein, Ann. d. Phys. **17**, 132 (1905)

[28] G. I. Taylor, Proc. Cam. phil. Soc. **15**, 114 (1909).

[29] P. A. M. Dirac, *The Principles of Quantum Mechanics*, 4th edition, Clarendon Press, London (1958).

[30] R. Wiegner, *A new source of entangled photons: spatial and temporal correlations in the fluorescence light of trapped atoms*, Diploma thesis, University of Erlangen-Nuremberg (2008).

[31] S. J. van Enk, Phys. Rev. A **72** 064306 (2005).

[32] S. J. van Enk, Phys. Rev. A **74** 026302 (2006).

[33] A. N. Boto et al., Rev. Lett. **85**, 2733 (2000).

[34] C. Santori et al., Nature (London) **419**, 594 (2002).

[35] C. Hettich et al., Science **298**, 385 (2002).

[36] T. Legero et al., Phys. Rev. Lett. **93**, 070503 (2004).

[37] J. Beugnon et al., Nature (London) **440**, 779 (2006).

[38] P. Maunz et al., Nature Phys. **3**, 538 (2007).

[39] F. Dubin et al., Phys. Rev. Lett. **98**, 183003 (2007).

[40] F. Dubin et al., Phys. Rev. Lett. **99**, 183001 (2007).

[41] Q. A. Turchette et al., Phys. Rev. Lett. **81**, 3631 (1998).

[42] C. Roos et al., Phys. Rev. Lett. **83**, 4713 (1999).

[43] B. Darquié et al., Science **309**, 454 (2005).

[44] J. Volz et al., Phys. Rev. Lett. **96**, 030404 (2006).

[45] T. Basché et al., Phys. Rev. Lett. **69**, 1516 (1992).

[46] B. Lounis and W. E. Moerner, Nature (London) **407**, 491 (2000).

[47] G. Wrigge et al., Nature Phys. **4**, 60 (2008).

[48] A. Gruber et al., Science **276**, 2012 (1997).

[49] R. Brouri et al., Opt. Lett. **25**, 1294 (2000).

[50] C. Kurtsiefer et al., Phys. Rev. Lett. **85**, 290 (2000).

[51] P. Michler et al., Science **290**, 2281 (2000).

[52] V. Zwiller et al., Appl. Phys. Lett. **78**, 2476 (2001).

[53] H. Dehmelt, in *Advances in Atomic and Molecular Physics*, Vol. **3**, Academic Press Inc., New York (1969); Vol. **5**, Academic Press Inc., New York (1969).

[54] R. C. Thompson, in *Advances in Atomic, Molecular and Optical Physics*, Vol. **30**, Academic Press Inc., Boston (1993).

[55] G. S. Agarwal, Quantum Optics, vol. 70 of Springer Tracts in Modern Physics (Springer, Berlin, 1974).

[56] R. H. Lehmberg, Phys. Rev. A **2**, 883(1970).

[57] R. J. Glauber, Phys. Rev. **130**, 2529 (1963).

[58] R. J. Glauber, Phys. Rev. **131**, 2766 (1963).

[59] Z. Y. Ou, Phys. Rev. A **37**, 1607 (1988).

[60] G. S. Agarwal et al., Phys. Rev. A **65**, 053826 (2002).

[61] C. Skornia et al., Phys. Rev. A **64**, 063801 (2001).

[62] J. von Zanthier, T. Bastin, and G. S. Agarwal, Phys. Rev. A **74**, 061802(R) (2006).

[63] C. Thiel, Diploma thesis, University of Erlangen-Nuremberg (2006).

[64] R. H. Dicke, Phys. Rev. **93**, 99 (1954).

[65] G. S. Agarwal, *Quantum Optics, Springer Tracts in Modern Physics* (Springer, Berlin, 1974), Vol. 70.

[66] J. S. Bell, Physics **1**, 195 (1964).

[67] J. F. Clauser and M. A. Horne, Phys. Rev. D **10**, 526 (1974).

[68] J. F. Clauser, M. A. Horne, A. Shimony, and R. A. Holt, Phys. Rev. Lett. **23**, 880 (1969).

[69] N. Kiesel et al., Phys. Rev. Lett. **98**, 063604 (2007).

[70] T. B. Pittman et al., Phys. Rev. A **51**, 3495 (1995).

[71] P. Walther et al., Nature (London) **429**, 158 (2004).

[72] M. W. Mitchell, J. S. Lundeen, and A. M. Steinberg, Nature (London) **429**, 161 (2004).

[73] M. D'Angelo, C. V. Chekhova, and Y. Shih, Phys. Rev. Lett. **87**, 013602 (2001).

[74] P. R. Hemmer et al., Phys. Rev. Lett. **96**, 163603 (2006).

[75] M. C. Teich, B. E. A. Saleh, Československý Časopis pro Fyziku (Prague) **4**, 73 (1997).

[76] A. Muthukrishnan, M. O. Scully, and M. S. Zubairy, J. Opt. B **6**, S575 (2004).

[77] G. S. Agarwal et al., Phys. Rev. A **70**, 063816 (2004).

[78] L. A. Lugiato, A. Gatti, and E. Brambilla, J. Opt. B: Quantum Semiclass. Opt. **4**, S176 (2002).

[79] Y. Shih, IEEE J. Sel. Topics Quantum Electron. **13**, 1016 (2007).

[80] Lord Rayleigh, Philos. Mag. **8**, 261 (1879).

[81] E. Abbe, M. Schultzes Archiv f. mikr. Anat. **9**, 413 (1873); Ges. Abh. I, 45 (1904).

[82] E. J. S. Fonseca, C. H. Monken, and S. Pàdua, Phys. Rev. Lett. **82**, 2868 (1999).

[83] M. D'Angelo, M. V. Chekhova, and Y. Shih, Phys. Rev. Lett. **87**, 013602 (2001).

[84] A. F. Abouraddy et al., J. Opt. B: Quantum Semiclass. Opt. **3**, 50 (2001).

[85] K. Edamatsu, R. Shimizu, and T. Itoh, Phys. Rev. Lett. **89**, 213 601 (2002).

[86] R. S. Bennink, S. J. Bentley, R. W. Boyd, and J. C. Howell, Phys. Rev. Lett. **92**, 033601 (2004).

[87] H. Lee, P. Kok, and J. P. Dowling, J. Mod. Opt. **49**, 2325 (2002).

[88] J. J. Bollinger, W. M. Itano, D. J. Wineland, and D. J. Heinzen, Phys. Rev. A **54**, R4649 (1996).

[89] G. S. Agarwal and M. O. Scully, Opt. Lett. **28**, 462 (2003).

[90] U. W. Rathe and M. O. Scully, Lett. Math. Phys. **34**, 297 (1995).

[91] R. J. Glauber, M. Kleber, A. K. Patnaik, M. O. Scully, and H. Walther, J. Phys. B **38**, S521 (2005).

[92] L. Mandel, Phys. Rev. A **28**, 929 (1983).

[93] Y. Shih, Eur. Phys. J. D **22**, 485 (2003).

[94] U. Eichmann et al., Phys. Rev. Lett. **70**, 2359 (1993).

[95] Y. Miroshnychenko et al., Nature **442**, 151 (2006).

[96] D. Stick et al., Nature Physics **2**, 36 (2006).

[97] A. F. Abouraddy et al., Phys. Rev. Lett. **87**, 123602 (2001).

[98] M. Born and E. Wolf, *Principles of Optics*, 7th edition (Cambridge University Press, Cambridge, 1999).

[99] G. Scarcelli, A. Valencia, and Y. Shih, Europhys. Lett. **68**, 618 (2004).

[100] B. B. Blinov et al., Nature (London) **428**, 153 (2004).

[101] W. K. Wootters, Phys. Rev. Lett. **80**, 2245 (1998).

[102] W. Dür, G. Vidal, and J. I. Cirac, Phys. Rev. A **62**, 062314 (2000).

[103] F. Verstraete et al., Phys. Rev. A **65**, 052112 (2002).

[104] L. Lamata, J. León, D. Salgado, and E. Solano, Phys. Rev. A **75**, 022318 (2007).

[105] D. Bouwmeester et al., Phys. Rev. Lett. **82**, 1345 (1999).

[106] J.-W. Pan et al., Phys. Rev. Lett. **86**, 4435 (2001).

[107] Z. Zhao et al., Nature **430**, 54 (2004).

[108] C.-Y. Lu et al., Nature Phys. **3**, 91 (2007).

[109] W. Wieczorek et al., e-print arXiv: quant-ph/09032213 (2009).

[110] R. Prevedel et al., e-print arXiv: quant-ph/09032212 (2009).

[111] B. Julsgaard, A. Kozhekin, and E. S. Polzik, Nature (London) **413**, 400 (2001).

[112] D. N. Matsukevich et al., Phys. Rev. Lett. **96**, 030405 (2006).

[113] C. W. Chou et al., Nature (London) **438**, 828 (2005).

[114] P. Maunz et al., e-print arXiv: quant-ph/0608047.

[115] J. K. Stockton et al., Phys. Rev. A **67**, 022112 (2003).

[116] M. Bourennane et al., Phys. Rev. Lett. **96**, 100502 (2006).

[117] G. Tóth, J. Opt. Soc. Am. B **24**, 275 (2007).

[118] A.R. Usha Devi, R. Prabhu, and A.K. Rajagopal, Phys. Rev. Lett. **98**, 060501 (2007).

[119] A. Retzker, E. Solano, and B. Reznik, Phys. Rev. A **75**, 022312 (2007).

[120] L. Mandel and E.Wolf, *Quantum coherence and quantum optics* (Cambridge University Press, Cambridge, 1995).

[121] F. Tamburini, B. A. Bassett, and C. Ungarelli, Phys. Rev. A **78**, 012114 (2008).

[122] E. P. Wigner, *Group Theory and its application to the quantum mechanics of atomic spectra* (Academic Press, New York, 1959).

[123] K. Hagiwara et al., Phys. Rev. D **66**, 010001 (2002).

[124] A. Acín et al., Phys. Rev. Lett. **87**, 040401 (2001).

[125] L. Chen and Y.-X. Chen, Phys. Rev. A **74**, 062310 (2006).

[126] L. Lamata et al., Phys. Rev. A **74**, 052336 (2007).

[127] D. M. Greenberger et al., Am. J. Phys. **58**, 1131 (1990).

[128] A. Rauschenbeutel et al., Science **288**, 2024 (2000).

[129] J.-W. Pan et al., Nature **403**, 515 (2000).

[130] B. P. Lanyon and N. K. Langford, New J. Phys. **11**, 013008 (2009).

[131] C. A. Sackett et al., Nature **404**, 256 (2000).

[132] D. Leibfried et al., Science **304**, 1476 (2004).

[133] P. Walther, K. J. Resch, and A. Zeilinger, Phys. Rev. Lett. **94**, 240501 (2005).

[134] V. Coffman, J. Kundu, and W. K. Wootters, Phys. Rev. A **61**, 052306 (2000).

[135] S. Freedman and J. F. Clauser, Phys. Rev. Lett. **28**, 938 (1972).

[136] J. F. Clauser, Phys. Rev. Lett. **36**, 1223 (1976).

[137] E. S. Fry and R. C. Thompson, Phys. Rev. Lett. **37**, 465 (1976).

[138] N. D. Mermin, Phys. Rev. Lett. **65**, 1838 (1990).

[139] R. F. Werner and M. M. Wolf, Phys. Rev. A **64**, 032112 (2001).

[140] W. Son, J. Lee, and M. S. Kim, Phys. Rev. Lett. **96**, 060406 (2006).

[141] D. Bohm, *Quantum Theory*, Prentice-Hall, Inc., New York (1951).

[142] D. Bohm and Y. Aharonov, Phys. Rev. **108**, 1070 (1957).

[143] D. N. Matsukevich, Phys. Rev. Lett. **100**, 150404 (2008).

[144] D. Göring, *Quantum imaging using second order correlations in the fluorescense light of atoms*, University of Erlangen-Nuremberg (2007); available at http://www.enb.physik.uni-erlangen.de/includes/filerouter.php?id=143

[145] I. Pitowsky, *Quantum Probability Quantum Logic*, Vol. 321 of Lecture Notes in Physics (Springer, Berlin 1989).

[146] S. Janssens, B. De Baets, and H. De Meyer, Fuzzy Sets Syst. **148**, 263 (2004).

[147] R. F. Werner, Phys. Rev. A **40**, 4277 (1989).

[148] S. Popescu, Phys. Rev. Lett. **74**, 2619 (1995).

[149] S. Teufel *et al.*, Phys. Rev. A **56**, 1217 (1997).

[150] J. Barrett, Phys. Rev. A **65**, 042302 (2002).

[151] A. Acín, N Gisin, and B. Toner, Phys. Rev. A **73**, 062105 (2006).

[152] N. Gisin, e-print arXiv: quant-ph/0702021.

[153] C. F. Wildfeuer and J. D. Dowling, Phys. Rev. A **78**, 032113 (2008).

[154] L. Dunningham and V. Vedral, Phys. Rev. Lett. **99**, 180404 (2007).

I want morebooks!

Buy your books fast and straightforward online - at one of the world's fastest growing online book stores! Environmentally sound due to Print-on-Demand technologies.

Buy your books online at
www.get-morebooks.com

Kaufen Sie Ihre Bücher schnell und unkompliziert online – auf einer der am schnellsten wachsenden Buchhandelsplattformen weltweit!
Dank Print-On-Demand umwelt- und ressourcenschonend produziert.

Bücher schneller online kaufen
www.morebooks.de

OmniScriptum Marketing DEU GmbH
Heinrich-Böcking-Str. 6-8
D - 66121 Saarbrücken
Telefax: +49 681 93 81 567-9

info@omniscriptum.com
www.omniscriptum.com

Printed by Books on Demand GmbH, Norderstedt / Germany